McGraw-Hill Education

Trigonometry
Review and Workbook

McGraw-Hill Education

Trigonometry
Review and Workbook

William D. Clark, PhD

Sandra Luna McCune, PhD

New York Chicago San Francisco Athens London Madrid
Mexico City Milan New Delhi Singapore Sydney Toronto

3 4 5 6 7 8 9   LHS   24 23 22 21 20

ISBN   978-1-260-12892-5
MHID   1-260-12892-X

e-ISBN   978-1-260-12893-2
e-MHID   0-1-260-12893-8

McGraw-Hill books are available at special quantity discounts to use as premiums and sales promotions, or for use in corporate training programs. To contact a representative, please visit the Contact Us page at www.mhprofessional.com.

Contents

Introduction   xi

CHAPTER 1

Angles and Their Measure   1

Definitions and Terminology   1
Complementary and Supplementary Angles   4
Coterminal Angles and Reference Angles   4
Radian Measure   7

CHAPTER 2

Concepts from Geometry   11

The Sum of a Triangle's Angles and the Triangle Inequality   11
The Pythagorean Theorem   14

CHAPTER 3

Right Triangle Trigonometry   19

Trigonometric Ratios of an Acute Angle
in a Right Triangle   19
Trigonometric Ratios of Special Acute Angles   22

CHAPTER 4

General Right Triangles   25

Solving Right Triangles   25
Applications of Right Triangle Trigonometry   27

CHAPTER 5

Oblique Triangles 31

Law of Cosines (SAS or SSS) 31

Law of Sines (ASA or AAS) 35

Law of Sines Ambiguous Case (SSA) 39

Solving General Triangles 41

Area of a General Triangle Using Trigonometry 46

CHAPTER 6

Trigonometric Functions of Any Angle 49

Definitions of the Trigonometric Functions 49

Trigonometric Functions of Complementary Angles 53

The Unit Circle 56

Trigonometric Functions of Quadrantal Angles 59

Trigonometric Functions of Coterminal Angles 61

Trigonometric Functions of Negative Angles 63

Using Reference Angles to Find the Values of
Trigonometric Functions 64

CHAPTER 7

Trigonometric Identities 69

Definition and Guidelines 69

The Reciprocal and Ratio Identities 71

The Pythagorean Identities 72

Sum and Difference Formulas for
the Sine Function 74

Sum and Difference Formulas for
the Cosine Function 76

Sum and Difference Formulas for the Tangent Function 78

Reduction Formulas 80

Double-Angle Identities 81

Half-Angle Identities 83

Sum-to-Product Identities 85

Product-to-Sum Identities 86

CHAPTER 8

Trigonometric Functions of Real Numbers 87

Definitions and Basic Concepts of Trigonometric Functions of Real Numbers 87

Periodic Functions 89

CHAPTER 9

Graphs of the Sine Function 93

The Graph of $y = \sin x$ 93

The Graph of $y = A \sin x$ 95

The Graph of $y = A \sin Bx$ 97

The Graph of $y = A \sin (Bx - C)$ 100

The Graph of $y = A \sin (Bx - C) + D$ 103

CHAPTER 10

Graphs of the Cosine Function 107

The Graph of $y = \cos x$ 107

The Graph of $y = A \cos (Bx - C) + D$ 109

CHAPTER 11

Graphs of the Tangent Function 113

The Graph of $y = \tan x$ 113

The Graph of $y = A \tan (Bx - C) + D$ 114

CHAPTER 12

Graphs of the Secant, Cosecant, and Cotangent Functions 117

The Graph of $y = A \sec (Bx - C) + D$ 117

The Graph of $y = A \csc (Bx - C) + D$ 119

The Graph of $y = A \cot (Bx - C) + D$ 121

CHAPTER 13

Inverse Trigonometric Functions 125

The Inverse Sine, Cosine, and Tangent Functions 125

The Inverse Secant, Cosecant, and Cotangent Functions 130

CHAPTER 14

Solving Trigonometric Equations 135

Basic Concepts of Trigonometric Equations 135

Solving for Exact Solutions to Trigonometric Equations 139

Solving for Approximate Solutions to Trigonometric Equations 141

CHAPTER 15

Trigonometric Form of a Complex Number 143

Definition of the Trigonometric Form of a Complex Number 143

The Product and Quotient of Trigonometric Forms of Complex Numbers 145

De Moivre's Theorem 148

Roots of Complex Numbers 149

CHAPTER 16

Polar Coordinates 153

Basic Concepts of Polar Coordinates 153

Converting Between Coordinate Systems 155

Graphing Equations in Polar Form 157

GLOSSARY

Glossary   163

APPENDIX A

Calculator Instructions for Trigonometry Using the TI-84 Plus   169

General Usage   169
Setting the Calculator to Degree or Radian Mode   170
Overriding Radian or Degree Mode   170
Evaluating Trigonometric Functions   170
Determining Inverse Trigonometric Values   173
Graphing Polar Equations   176

APPENDIX B

Trigonometric Identities   179

APPENDIX C

The Complex Plane   181

Answer Key   183

Introduction

Welcome to *Trigonometry Review and Workbook!* In this book, you will find an interactive, student-friendly approach that leads to mastery of trigonometry. Each lesson features illustrative examples followed by an abundant number of exercises for practice. In addition, sidebars in the margins contain cautions against common errors and other helpful advice.

The book takes you from right and oblique triangle trigonometry, including real-world applications, to concepts of trigonometric functions and identities, and ends with polar coordinates. The topics are aligned with the Common Core State Standards (CCSS), which outline a set of high-quality academic standards in mathematics. These standards have been adopted by over 40 states, the District of Columbia, and four territories of the United States. Likewise, the topics cover the essential math curricula of non-CCSS states and Canada.

Essential components of learning trigonometry are active participation and lots of practice. Accordingly, this book is designed as a workbook with more than 1,000 problems to help you solidify your trigonometric skills and understanding. Take your time to read the lessons and work the practice problems at a comfortable pace.

No matter how much it might seem that trigonometry is too hard to master, you can overcome this by building your confidence in your ability to work with trigonometric principles and models. Along the way, don't be afraid of making mistakes. Mistakes are part of the learning process. Learn to view them as opportunities for growth and a deeper understanding of concepts. At the end of the day, success will come.

Trigonometry Review and Workbook is written in the hope of sharing the excitement of becoming competent and sure as a student of trigonometry. The goal of the book is to help you achieve that feeling of accomplishment!

CHAPTER 1

Angles and Their Measure

Definitions and Terminology

A **ray** is a half-line beginning at an endpoint and extending indefinitely in one direction from that point. An **angle** is formed by two rays that have a common endpoint. The common endpoint is the **vertex**. The angle begins with the two rays lying on top of one another. One ray, the **initial side**, is fixed in place, and the other ray, the **terminal side**, is rotated about the vertex. Indicate the rotation with a small arrow close to the vertex.

Angle θ

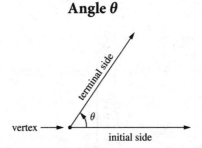

An angle in **standard position** has its vertex located at the origin, and its initial side extends along the positive x-axis. The terminal side can rotate in either a counterclockwise or a clockwise direction and can lie in any quadrant or on any axis. When the terminal side lies in a quadrant, then you say the angle lies in that quadrant. If the terminal side lies on an axis, the angle is a **quadrantal angle**.

Greek letters, for example, α (alpha), β (beta), γ (gamma), θ (theta), and ϕ (phi), as well as uppercase Roman letters, such as A, B, and C, are used to label angles.

For convenience, in figures, quadrant I is abbreviated as I, quadrant II is abbreviated as II, and so forth.

▶ θ lies in quadrant II:

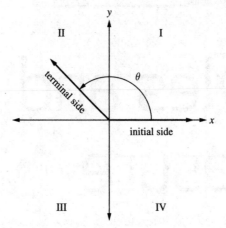

▶ α lies in quadrant III:

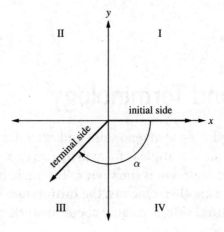

▶ β is a quadrantal angle:

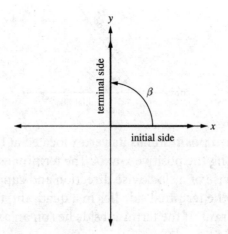

The **measure of an angle** is the amount of rotation from the initial side to the terminal side. *Counterclockwise* rotations yield positive angles, and *clockwise* rotations yield negative angles.

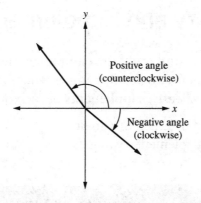

Positive angle
(counterclockwise)

Negative angle
(clockwise)

> The measure of an angle depends on the amount of rotation, not on the perceived length of the rays forming the angle.

You are likely familiar with measuring angles using degrees. One **degree** (1°) is $\frac{1}{360}$ of a full revolution (counterclockwise) about the vertex. So one complete counterclockwise revolution corresponds to 360°.

Recall from elementary geometry that an angle is **acute** if it is between 0° and 90°, a **right angle** if it equals 90°, an **obtuse angle** if it is between 90° and 180°, a **straight angle** if it equals 180°, and a **reflex angle** if it is between 180° and 360°:

15°	→	acute angle
300°	→	reflex angle
150°	→	obtuse angle

> - It is common to use the label for an angle to refer to both the angle and its measure.
> - The degree measure of an angle should include the word *degree* or the degree symbol ° after the number; e.g., .45 degrees = 45°.

EXERCISE 1-1

For questions 1 to 10, state the quadrant in which the angle lies or the axis on which it lies.

1. 20°

2. 180°

3. 97°

4. 236°

5. −18°

6. 121°

7. 270°

8. 54°

9. −143°

10. −235°

For questions 11 to 20, determine whether the statement is True or False.

11. 35° is an acute angle.

12. 120° is a reflex angle.

13. 250° is an obtuse angle.

14. 89° is an obtuse angle.

15. −90° is a quadrantal angle.

16. −45° is an acute angle.

17. 91° is an obtuse angle.

18. 270° is a quadrantal angle.

19. 0° is a straight angle.

20. All right angles have the same measure.

Complementary and Supplementary Angles

Two positive angles whose sum is 90° are **complementary**; or, equivalently, **complements** of each other. Two positive angles whose sum is 180° are **supplementary**; or, equivalently, **supplements** of each other.

30° and 60°　→　complementary angles
45° and 135°　→　supplementary angles

EXERCISE 1-2

For questions 1 to 10, find the complement of the given angle.

1. 40°　　　　　　　　　　　　　**6.** 21°

2. 16°　　　　　　　　　　　　　**7.** 86°

3. 88°　　　　　　　　　　　　　**8.** 61°

4. 54°　　　　　　　　　　　　　**9.** 20°

5. 47°　　　　　　　　　　　　　**10.** 13°

For questions 11 to 20, find the supplement of the given angle.

11. 120°　　　　　　　　　　　　**16.** 60°

12. 22°　　　　　　　　　　　　　**17.** 166°

13. 81°　　　　　　　　　　　　　**18.** 3°

14. 90°　　　　　　　　　　　　　**19.** 143°

15. 101°　　　　　　　　　　　　**20.** 18°

Coterminal Angles and Reference Angles

Coterminal angles have the same initial and terminal sides. To find an angle that is coterminal to a given angle, add or subtract 360°. In general, a given angle θ is coterminal with $\theta + n(360°)$, where n is a nonzero integer. If an angle is greater than 360° or is negative, you can find an equivalent nonnegative coterminal angle that is less than 360° by adding or subtracting a positive integer multiple of 360°:

0° and 360°　　→　coterminal

45° and 405°　　→　coterminal

−30° and 330°　　→　coterminal

800° and 80°　　→　coterminal

The **reference angle** for a non-quadrantal angle in standard position is the acute angle formed by the terminal side of the angle and the *x*-axis.

The following are examples of an angle θ and its reference angle θ_r:

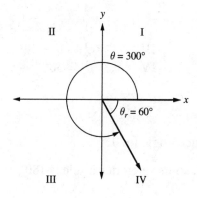

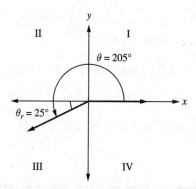

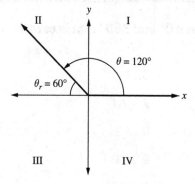

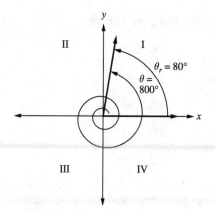

The relationship of a positive angle θ that is less than $360°$ and its reference angle θ_r in each quadrant is given in the following table.

θ's Quadrant Relationship

Quadrant	Relationship
I	$\theta_r = \theta$
II	$\theta_r = 180° - \theta$
III	$\theta_r = \theta - 180°$
IV	$\theta_r = 360° - \theta$

EXAMPLE

▶ Determine the reference angle for $135°$.

$135°$ is in quadrant II, so its reference angle is $180° - 135° = 45°$.

EXAMPLE

▶ Determine the reference angle for $930°$.

Because $930° > 360°$, find its coterminal angle less than $360°$ first.

$930° - (2 \times 360°) = 930° - 720° = 210°$.

$210°$ is in quadrant III, so its reference angle is $210° - 180° = 30°$.

EXERCISE 1-3

For questions 1 to 10, determine an angle between $0°$ and $360°$ that is coterminal with the given angle and specify its quadrant.

1. $-17°$

2. $380°$

3. $466°$

4. $-101°$

5. $-14°$

6. $700°$

7. $564°$

8. $-293°$

9. $-314°$

10. $902°$

For questions 11 to 20, find the reference angle for the given angle.

11. $130°$

12. $230°$

13. $345°$

14. $95°$

15. $117°$

16. $400°$

17. $-17°$

18. $268°$

19. $-195°$

20. $1045°$

Radian Measure

A **radian** is the measure of a central angle of a circle that intercepts an arc whose length equals the radius of that circle.

Because the circle's circumference equals 2π times the radius, a full circular rotation is 2π radians. Therefore, you have the following equivalencies:

A **central angle** is an angle formed at the center of a circle by two radii.

▶ 2π radians $= 360°$

▶ π radians $= 180°$

▶ 1 radian $= \dfrac{180°}{\pi} \ (\approx 57.3°)$

▶ $\dfrac{\pi}{180}$ radians $= 1°$

Therefore, to convert degrees to radians, multiply the degree measure by $\dfrac{\pi}{180°}$; and, conversely, to convert radians to degrees, multiply the radian measure by $\dfrac{180°}{\pi}$.

It is not necessary to write the units "radians" after a radian measure. You can assume that an angle measure that is not labeled with "degrees" or the degree symbol is in radians.

EXAMPLE

▶ Convert $60°$ to radians in terms of π.

$$60°\left(\frac{\pi}{180°}\right) = \cancel{60°}\left(\frac{\pi}{\cancel{180°}_{3}}\right) = \frac{\pi}{3}$$

EXAMPLE

▶ Convert $\dfrac{3\pi}{4}$ to degrees.

$$\frac{3\pi}{4}\left(\frac{180°}{\pi}\right) = \frac{3\cancel{\pi}}{\cancel{4}_{1}}\left(\frac{\cancel{180°}^{45°}}{\cancel{\pi}}\right) = 135°$$

EXAMPLE

▶ Convert $-\dfrac{\pi}{6}$ to degrees.

$$-\frac{\pi}{6}\left(\frac{180°}{\pi}\right) = -\frac{\cancel{\pi}}{\cancel{6}_{1}}\left(\frac{\cancel{180°}^{30°}}{\cancel{\pi}}\right) = -30°$$

Convert an angle measure of 3 to degrees (round your answer to the nearest tenth of a degree).

$$3\left(\frac{180°}{\pi}\right) = \frac{540°}{\pi} \approx 171.9°$$

Following are some applications of radian measure of angles.

Arc length: On a circle of radius r, a central angle θ (measured in radians) intercepts an arc of length $s = r\theta$.

Find the length of an arc intercepted by a central angle of $\frac{5\pi}{6}$ in a circle of radius 24 inches (see figure below). (Round your answer to one decimal place.)

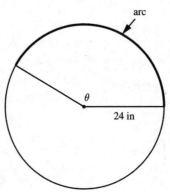

$$s = r\theta = \left(24 \text{ in}\right)\left(\frac{5\pi}{6}\right) = 20\pi \text{ in} \approx 62.8 \text{ in}$$

Find the length of an arc intercepted by a central angle of 315° in a circle of radius 36 centimeters. (Round your answer to one decimal place.)

First, $315°\left(\frac{\pi}{180°}\right) = \frac{7\pi}{4}$; then $s = r\theta = \left(36 \text{ cm}\right)\left(\frac{7\pi}{4}\right) = 63\pi \text{ cm} \approx 197.9 \text{ cm}$

A **sector** is the portion of the interior of a circle enclosed by two radii and an arc of the circle.

Area of a sector: The area of a sector of a circle of radius r and central angle θ (measured in radians) is $A = \frac{1}{2}r^2\theta$.

EXAMPLE

Find the area of a sector intercepted by a central angle of $\dfrac{5\pi}{6}$ in a circle of radius 24 inches (see figure below). (Round your answer to one decimal place.)

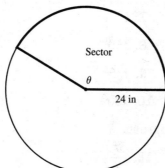

$$A = \frac{1}{2}r^2\theta = \frac{1}{2}(24 \text{ in})^2\left(\frac{5\pi}{6}\right) \approx 754.0 \text{ in}^2$$

EXAMPLE

Find the area of a sector intercepted by a central angle of 315° in a circle of radius 36 centimeters. (Round your answer to one decimal place.)

First, $315°\left(\dfrac{\pi}{180°}\right) = \dfrac{7\pi}{4}$; then $A = \dfrac{1}{2}r^2\theta = \dfrac{1}{2}(36 \text{ cm})^2\left(\dfrac{7\pi}{4}\right) \approx 3{,}562.6 \text{ cm}^2$

Angular speed: The angular speed of an object moving at a constant speed along a circular arc is $\dfrac{\theta}{t}$, where θ is the central angle measured in radians.

> **Angular speed** is the angle through which a rotating object travels per unit of time.

EXAMPLE

An engine is rotating at 540 revolutions per minute. Calculate its angular speed in radians per second.

Given that 1 revolution is 2π radians and 1 minute is 60 seconds, then

540 revolutions per minute in radians per second is $\dfrac{(540)(2\pi)}{60 \text{ sec}} = 18\pi$ radians per second.

EXERCISE 1-4

For questions 1 to 10, convert the given angle to radians in terms of π.

1. $30°$

2. $120°$

3. $45°$

4. $90°$

5. $270°$

6. $180°$

7. $1°$

8. $360°$

9. $135°$

10. $300°$

For questions 11 to 20, convert the given angle to degrees.

11. 3π

12. $\dfrac{5\pi}{4}$

13. $\dfrac{\pi}{2}$

14. $-\dfrac{\pi}{4}$

15. $\dfrac{5\pi}{6}$

16. $\dfrac{5\pi}{3}$

17. $\dfrac{9\pi}{4}$

18. $\dfrac{\pi}{6}$

19. $-\dfrac{\pi}{3}$

20. $\dfrac{7\pi}{6}$

For questions 21 to 25, convert the radian measure to degrees. (Round answers to one decimal place, as needed).

21. 3.7

22. 1.4

23. 0.95

24. 2.3

25. 4.6

For questions 26 to 30, solve as indicated. (Round answers to one decimal place, as needed).

26. Find the length of an arc intercepted by a central angle of $\dfrac{\pi}{3}$ in a circle of radius 18 feet.

27. Find the length of an arc intercepted by a central angle of $225°$ in a circle of radius 12 meters.

28. Find the area of a sector intercepted by a central angle of $\dfrac{\pi}{3}$ in a circle of radius 18 feet.

29. Find the area of a sector intercepted by a central angle of $225°$ in a circle of radius 12 meters.

30. A wheel is turning at the rate of 960 revolutions per minute. Calculate its angular speed in radians per second.

CHAPTER 2

Concepts from Geometry

The Sum of a Triangle's Angles and the Triangle Inequality

The sum of the interior angles of a triangle is 180°.

EXAMPLE

▶ In triangle *ABC* shown, solve for *B*.

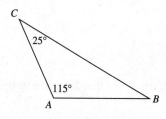

$$A + B + C = 180°$$
$$115° + B + 25° = 180°$$
$$B = 180° - 140°$$
$$B = 40°$$

Triangle inequality theorem   The sum of the lengths of any two sides of a triangle is greater than the length of the third side.

EXAMPLE

▶ Can the lengths 5, 8, and 11 be the lengths of the sides of a triangle? Yes or No?

▶ The longest side is 11. Compare $5 + 8$ and 11.

$5 + 8 > 11$?
$13 > 11$? True.

▶ The answer is Yes, 5, 8, and 11 can be the lengths of the sides of a triangle.

EXAMPLE

▶ Can the lengths 2, 6, and 9 be the lengths of the sides of a triangle? Yes or No?

▶ The longest side is 9. Compare $2 + 6$ and 9.

$2 + 6 > 9$?
$8 > 9$? False.

▶ The answer is No, 2, 6, and 9 cannot be the lengths of the sides of a triangle.

EXAMPLE

▶ Can the lengths 5, 12, and 7 be the lengths of the sides of a triangle? Yes or No?

▶ The longest side is 12. Compare $5 + 7$ and 12.

$5 + 7 > 12$?
$12 > 12$? False.

▶ The answer is No, 5, 12, and 7 cannot be the lengths of the sides of a triangle.

EXERCISE 2-1

For questions 1 to 5, state Yes or No as to whether the three angles can be the interior angles of a triangle.

1. $25°$, $75°$, $80°$

2. $60°$, $40°$, $90°$

3. $30°$, $110°$, $40°$

4. $50°$, $50°$, $50°$

5. $170°$, $5°$, $5°$

For questions 6 to 10, find the third interior angle of the triangle if the first two interior angles of the triangle are as given.

6. 70°, 30°

7. 100°, 20°

8. 55°, 83°

9. 24°, 61°

10. 95°, 31°

For questions 11 to 15, solve for the angle θ.

11.

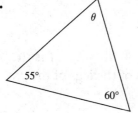

12.

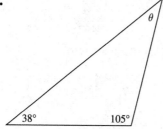

13.

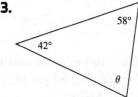

14.

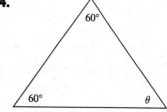

15.

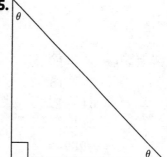

For questions 16 to 20, state Yes or No as to whether the given lengths can be the lengths of the sides of a triangle.

16. 8, 16, 22

17. 3, 4, 5

18. 8, 11, 2

19. 1, 1, 1

20. 502, 21, 485

The Pythagorean Theorem

In a right triangle, the **hypotenuse** is the side opposite the right angle, and the other two sides are the **legs** (illustrated below).

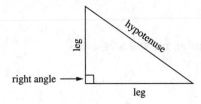

right angle ⟶

▶ **Pythagorean theorem** In a right triangle, $c^2 = a^2 + b^2$, where c is the length of the hypotenuse and a and b are the lengths of the legs of the right triangle.

EXAMPLE

▶ Find the length, c, of the hypotenuse of the right triangle shown.

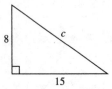

$$c^2 = a^2 + b^2$$
$$c^2 = 8^2 + 15^2$$
$$c^2 = 64 + 225$$
$$c^2 = 289$$
$$c = \sqrt{289}$$
$$c = 17$$

EXAMPLE

▶ Find b in the right triangle shown.

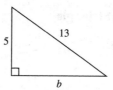

$$c^2 = a^2 + b^2$$
$$13^2 = 5^2 + b^2$$
$$169 = 25 + b^2$$
$$169 - 25 = 25 + b^2 - 25$$
$$144 = b^2$$
$$\sqrt{144} = b$$
$$12 = b$$

The Pythagorean theorem's converse is also true: In a triangle, if $c^2 = a^2 + b^2$, where c is the length of the triangle's longest side and a and b are the lengths of the triangle's other two sides, the triangle is a right triangle, and the right angle is opposite the longest side.

EXAMPLE

▶ Is a triangle with sides of lengths 3, 4, and 5 a right triangle? Yes or No?

▶ The longest side is 5. Compare 5^2 and $3^2 + 4^2$.

$5^2 = 3^2 + 4^2$?
$25 = 9 + 16$?
$25 = 25$? True.

The answer is Yes, a triangle with sides of lengths 3, 4, and 5 is a right triangle.

EXAMPLE

▶ Is a triangle with sides of length 6, 7, and 8 a right triangle? Yes or No?

▶ The longest side is 8. Compare 8^2 and $6^2 + 7^2$.

$8^2 = 6^2 + 7^2$?
$64 = 36 + 49$?
$64 = 85$? False.

The answer is No, a triangle with sides of lengths 6, 7, and 8 is not a right triangle.

EXERCISE 2-2

For questions 1 to 10, find the missing side of the right triangle.

1.
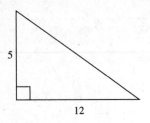

2.
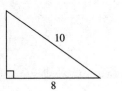

3.
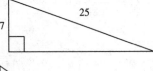

4.
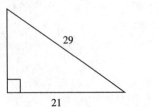

5.
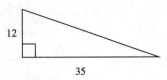

6.
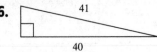

7.
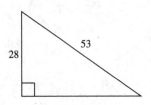

8.
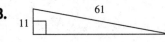

9.
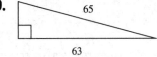

10.
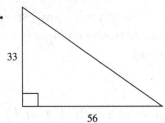

For questions 11 to 15, state Yes or No as to whether the set of numbers could be the lengths of the sides of a right triangle.

11. 13, 84, 85

12. 3, 4, 5

13. 6, 8, 22

14. 39, 80, 89

15. 17, 55, 41

For questions 16 to 20, solve as indicated. (Round answers to one decimal place, as needed.)

16. The lower end of a 25-foot pole that is leaning against a wall is 7 feet from the base of a building. At what height is the top of the pole touching the building?

17. An 18-foot lamp pole casts a shadow of 28 feet. What is the distance from the top of the pole to the tip of the shadow?

18. Find the length of the diagonal of a rectangular garden that has dimensions of 30 feet by 40 feet.

19. Two people want to carry a tall mirror that is a 9 by 9 feet square through a doorway that measures 3 feet by 8 feet. Will the mirror fit through the doorway?

20. Television sets are sized by the length of the diagonal of the face. Find the size of a portable TV whose rectangular face measures 27 inches by 21 inches.

CHAPTER 3

Right Triangle Trigonometry

Trigonometric Ratios of an Acute Angle in a Right Triangle

Consider a right triangle *ABC*, with the right angle at *C* and sides of lengths *a*, *b*, and *c*. The triangle's **hypotenuse** is $\overline{AB}$, the side opposite the right angle. Relative to the acute angle *A*, the leg $\overline{BC}$ is its **opposite** side, and the leg $\overline{AC}$ is its **adjacent** side (illustrated below).

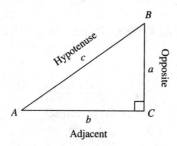

The six **trigonometric ratios** of angle *A* are defined as follows:

Name of Ratio	Abbreviation	Definition
sine A	sin A	$\sin A = \dfrac{\text{opposite}}{\text{hypotenuse}} = \dfrac{a}{c}$
cosine A	cos A	$\cos A = \dfrac{\text{adjacent}}{\text{hypotenuse}} = \dfrac{b}{c}$
tangent A	tan A	$\tan A = \dfrac{\text{opposite}}{\text{adjacent}} = \dfrac{a}{b}$
cosecant A	csc A	$\csc A = \dfrac{\text{hypotenuse}}{\text{opposite}} = \dfrac{c}{a}$
secant A	sec A	$\sec A = \dfrac{\text{hypotenuse}}{\text{adjacent}} = \dfrac{c}{b}$
cotangent A	cot A	$\cot A = \dfrac{\text{adjacent}}{\text{opposite}} = \dfrac{b}{a}$

Here is an example:

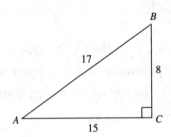

For the right triangle ABC shown above, the six trigonometric ratios for angle A are as follows:

A mnemonic for remembering the sine, cosine, and tangent ratios is "soh-cah-toa" (soh-kuh-toh-uh), formed from the first letters of "**s**ine is **o**pposite over **h**ypotenuse; **c**osine is **a**djacent over **h**ypotenuse; and **t**angent is **o**pposite over **a**djacent."

$$\sin A = \frac{\text{opposite}}{\text{hypotenuse}} = \frac{8}{17} \qquad \cos A = \frac{\text{adjacent}}{\text{hypotenuse}} = \frac{15}{17} \qquad \tan A = \frac{\text{opposite}}{\text{adjacent}} = \frac{8}{15}$$

$$\csc A = \frac{\text{hypotenuse}}{\text{opposite}} = \frac{17}{8} \qquad \sec A = \frac{\text{hypotenuse}}{\text{adjacent}} = \frac{17}{15} \qquad \cot A = \frac{\text{adjacent}}{\text{opposite}} = \frac{15}{8}$$

The values of the trigonometric ratios have no units, so they are pure numbers.

Observe that the pairs sin A and csc A, cos A and sec A, and tan A and cot A are reciprocals. Thus, you have the following **reciprocal relationships** given any angle θ:

Because of the reciprocal relationships, for the most part, you can focus on the more frequently used sine, cosine, and tangent ratios.

Reciprocal Relationships of Trigonometric Ratios

$$\csc\theta = \frac{1}{\sin\theta} \qquad \sec\theta = \frac{1}{\cos\theta} \qquad \cot\theta = \frac{1}{\tan\theta}$$

$$\sin\theta = \frac{1}{\csc\theta} \qquad \cos\theta = \frac{1}{\sec\theta} \qquad \tan\theta = \frac{1}{\cot\theta}$$

Here are examples:

Given $\sin\theta = \dfrac{3}{5}$, then $\csc\theta = \dfrac{5}{3}$.

Given $\cot\beta = \dfrac{5}{12}$, then $\tan\beta = \dfrac{12}{5}$.

EXERCISE 3-1

For questions 1 to 6, find the indicated trigonometric ratio for angle *R* using the right triangle below.

1. $\sin R$

2. $\cos R$

3. $\csc R$

4. $\cot R$

5. $\tan R$

6. $\sec R$

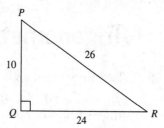

For questions 7 to 12, find the indicated trigonometric ratio for angle *B* using the right triangle below.

7. $\sin B$

8. $\sec B$

9. $\csc B$

10. $\cos B$

11. $\tan B$

12. $\cot B$

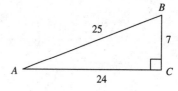

For questions 13 to 20, answer as indicated.

13. Given $\csc\theta = \dfrac{41}{9}$, find $\sin\theta$.

14. Given $\tan\beta = \dfrac{12}{35}$, find $\cot\beta$.

15. Given $\sec\gamma = \dfrac{61}{11}$, find $\cos\gamma$.

16. For an acute angle θ in a right triangle, the side opposite is 6 and the side adjacent is 8. Determine $\sin\theta$.

17. For an acute angle θ in a right triangle, the side opposite is 9 and the side adjacent is 40. Determine $\cos\theta$.

18. For an acute angle β in a right triangle, the side opposite is 12 and the side adjacent is 35. Determine $\cos\beta$.

19. Use the right triangle shown to determine $\tan\theta$.

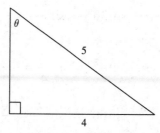

20. Use the right triangle shown to determine sec β.

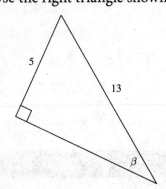

Trigonometric Ratios of Special Acute Angles

Two important right triangles are the $30°-60°-90°$ right triangle and the $45°-45°-90°$ right triangle. These two right triangles contain special acute angles (illustrated below).

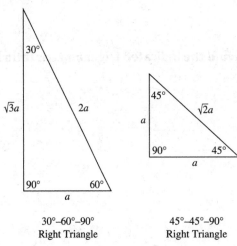

30°–60°–90°
Right Triangle

45°–45°–90°
Right Triangle

Because hereafter you will frequently encounter these special acute angles, you should learn to construct the right triangles that contain them.

Using the information shown in the figures, you can determine the following trigonometric ratios of the special acute angles 30°, 45°, and 60° $\left(\text{or } \dfrac{\pi}{6}, \dfrac{\pi}{4}, \text{ and } \dfrac{\pi}{3}, \text{ respectively}\right)$ associated with these two triangles:

Sines, Cosines, and Tangents of Special Acute Angles

$$\sin 30° = \sin\frac{\pi}{6} = \frac{1}{2} \quad \cos 30° = \cos\frac{\pi}{6} = \frac{\sqrt{3}}{2} \quad \tan 30° = \tan\frac{\pi}{6} = \frac{1}{\sqrt{3}} = \frac{\sqrt{3}}{3}$$

$$\sin 45° = \sin\frac{\pi}{4} = \frac{1}{\sqrt{2}} = \frac{\sqrt{2}}{2} \quad \cos 45° = \cos\frac{\pi}{4} = \frac{1}{\sqrt{2}} = \frac{\sqrt{2}}{2} \quad \tan 45° = \tan\frac{\pi}{4} = 1$$

$$\sin 60° = \sin\frac{\pi}{3} = \frac{\sqrt{3}}{2} \quad \cos 60° = \cos\frac{\pi}{3} = \frac{1}{2} \quad \tan 60° = \tan\frac{\pi}{3} = \sqrt{3}$$

All triangles similar to the two special triangles have the same trigonometric ratios as shown above. In general: **When determining the trigonometric ratios of an acute angle θ, you can use any right triangle that contains θ as one of its angles.**

In a given right triangle, if you are given one acute angle and the length of one side, you can use trigonometric ratios to determine missing side lengths in the triangle. For each missing side, select the trigonometric ratio that has the unknown side length as either the numerator or the denominator. Next, substitute the known and unknown information into the definition of the trigonometric ratio. Then solve for the missing side length.

EXAMPLE

▶ In triangle ABC with $C = 90°$, find a, given $A = 30°$ and $c = 48$.

▶ You are given an acute angle and the length of the hypotenuse. The unknown, a, is opposite the angle, so use the sine to determine a.

$\sin 30° = \dfrac{a}{48}; \dfrac{1}{2} = \dfrac{a}{48}$, which yields $a = 24$.

EXAMPLE

▶ In triangle RST with $S = 90°$, find r, given $R = 60°$ and $t = 5$. Write the exact answer in its simplest radical form.

▶ You are given an acute angle and the adjacent side. The unknown, r, is the opposite side, so use the tangent to determine r.

$\tan 60° = \dfrac{r}{5}; \sqrt{3} = \dfrac{r}{5}$, which yields $r = 5\sqrt{3}$.

EXERCISE 3-2

Solve for the exact length of the missing side in the right triangle. (Write the exact answer in simplest radical form for irrational answers.)

1. In triangle ABC with $C = 90°$, find b, given $A = 30°$ and $c = 26$.

2. In triangle ABC with $C = 90°$, find a, given $A = 60°$ and $c = 48$.

3. In triangle ABC with $C = 90°$, find c, given $B = 30°$ and $b = 20$.

4. In triangle ABC with $C = 90°$, find a, given $A = 45°$ and $c = 10$.

5. In triangle ABC with $C = 90°$, find b, given $A = 60°$ and $a = 9$.

6. In triangle ABC with $C = 90°$, find c, given $B = 45°$ and $b = 20$.

7. In triangle ABC with $C = 90°$, find b, given $A = 30°$ and $c = 108$.

8. In triangle ABC with $C = 90°$, find a, given $A = 60°$ and $c = 57$.

9. In triangle ABC with $C = 90°$, find b, given $A = 45°$ and $c = 16$.

10. In triangle ABC with $C = 90°$, find b, given $B = 30°$ and $a = 12$.

CHAPTER 4

General Right Triangles

Solving Right Triangles

This lesson (as in "Trigonometric Ratios of Special Acute Angles" in the previous chapter) involves solving for missing elements in right triangles. The difference in this lesson is that you will encounter angles other than the special angles introduced in the previous chapter. Therefore, you will find a calculator to be a valuable tool when you are working through the examples and exercises in this chapter. (See Appendix A for guidance in using the trigonometric features of the TI-84 Plus graphing calculator.)

A triangle has three sides and three angles. **Solving a triangle** means determining all of its angle measures and the lengths of all of its sides. In addition to knowing that a right triangle has one angle of 90°, you must either know the lengths of two sides or the length of one side and the measure of one acute angle in order to solve the triangle.

> When solving triangles, make sure your calculator is set to the desired mode of angle measurement (degrees or radians).

EXAMPLE

▶ Solve the triangle shown below.

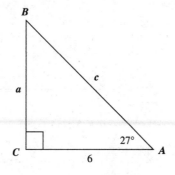

From the figure, you have $A = 27°$ and $b = 6$, the length of the side adjacent to A. Then, $B = 90° - 27° = 63°$. Use the tangent to find a, $\tan 27° = \dfrac{a}{6}$, which yields $a = 6\tan 27° \approx 3.1$. Use the cosine to find c,

$\cos 27° = \dfrac{6}{c}$, which yields $c = \dfrac{6}{\cos 27°} \approx 6.7$. Thus, the triangle is solved.

Solve the triangle shown below.

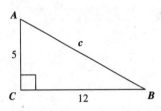

From the figure, you have $a = 12$ and $b = 5$, the lengths of the two legs of the right triangle. Then, $c^2 = 12^2 + 5^2 = 144 + 25 = 169$, from which you have $c = \sqrt{169} = 13$. For A, $\tan A = \dfrac{12}{5} = 2.4$, which yields $A \approx 67.4°$. (As shown in Appendix A, you determine A using the inverse tangent $\boxed{\text{2ND}}$ [tan−1] calculator key.) Then, $B \approx 90° - 67.4° = 22.6°$. Thus, the triangle is solved.

EXERCISE 4-1

Solve the given right triangle. (Round answers to one decimal place, as needed.)

1.
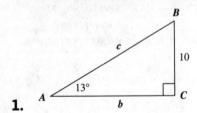

2.
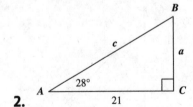

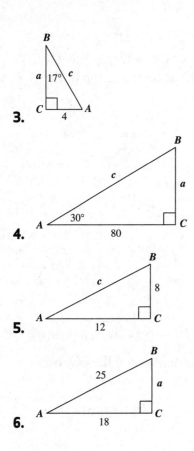

3.

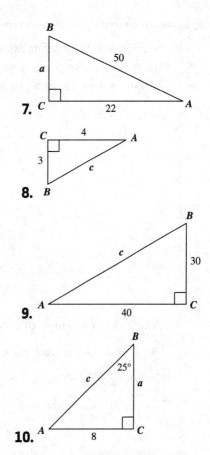

7.

8.

4.

5.

6.

9.

10.

Applications of Right Triangle Trigonometry

This lesson presents some of the various ways trigonometry can be applied in the real world. At the outset, among the concepts you will encounter are the angle of elevation and the angle of depression. The term **angle of elevation** refers to the angle from the horizontal upward to a point of interest, while the **angle of depression** refers to the downward angle formed between the horizontal and the line of sight to a point of interest below the horizontal (see illustration).

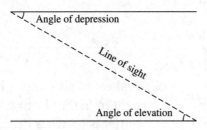

EXAMPLE

A 25-foot ladder leans against a house. If the foot of the ladder is 7 feet from the side of the house, what is the angle of elevation to the top of the ladder as seen from a point on the ground?

Make a sketch.

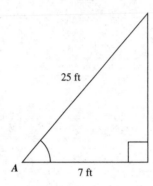

Angle A is the angle of elevation. Then $\cos A = \dfrac{7}{25} = 0.28$, which yields $A \approx 73.7°$ (As shown in Appendix A, you determine A using the inverse cosine $\boxed{2ND}\left[\cos^{-1}\right]$ calculator key.)

EXERCISE 4-2

Solve as indicated. (Round answers to one decimal place, as needed.)

1. A 40-foot support wire on a TV dish satellite makes an angle of 60° with the ground. Calculate the vertical distance above the ground of the point where the wire touches the dish.

2. A high-dive platform 80 feet off the ground is supported by a guy wire 140 feet long that is attached to the ground to support the platform. Calculate the measure of the angle where the wire touches the ground.

3. A forester wants to estimate the height of a giant tree from a distance of 150 feet. The angle of elevation from the forester's position to the top of the tree is 52°. Determine the height of the tree.

4. A ham radio antenna stands on top of a house. The base of the antenna is 20 feet from the ground. The angle of depression from the

top of the antenna to a point 100 feet from the base of the house measures 15°. How tall is the antenna? (See the diagram below.)

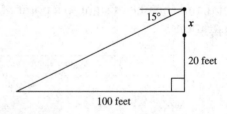

5. A sheet of aluminum is 20 inches wide. It is bent in half to form a V-shaped gutter. If the gutter is 6 inches deep, find the measure of the angle between the sides of the gutter.

6. A 6-foot-tall man casts a 9-foot shadow. What is the measure of the angle of elevation from the tip of the shadow on the ground to the top of the man's head?

7. A kite is attached to a string that is 200 feet long. If the string makes an angle of 40° with the ground, how high is the kite?

8. Each blade of a pair of scissors is 5 inches long from the pivot point of the scissors. When the scissors are opened so that the scissor points are 3 inches apart, what is the measure of the angle that the blades make with each other?

9. A solar panel is placed on a roof as shown in the figure below. What is the measure of the angle at point A?

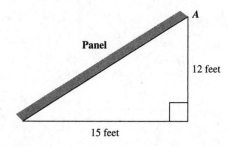

10. A nose cone on a rocket has cross-sectional measurements as shown below. How wide is the rocket that uses this nose cone?

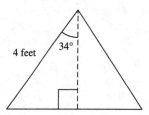

CHAPTER 5

Oblique Triangles

Law of Cosines (SAS or SSS)

The **Law of Cosines** consists of formulas for solving oblique triangles when you are given the lengths of two sides and the included angle between the two sides (abbreviated SAS) or the lengths of all three sides (abbreviated SSS) of a triangle.

▶ **The Law of Cosines**

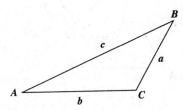

> An **oblique** triangle contains no right angle.

For the triangle shown, the Law of Cosines is as follows:

$$a^2 = b^2 + c^2 - 2bc \cos A$$
$$b^2 = a^2 + c^2 - 2ac \cos B$$

$$c^2 = a^2 + b^2 - 2ab \cos C$$

Note that the Law of Cosines formulas also can be used when you have a right triangle. In point of fact, the Law of Cosines is a generalization of the Pythagorean theorem.

EXAMPLE

▶ Find *c* as shown in the figure.

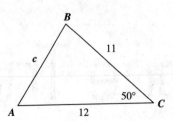

▶ Substitute the information from the figure into $c^2 = a^2 + b^2 - 2ab\cos C$ and then solve for *c*.

$$c^2 = 11^2 + 12^2 - 2(11)(12)\cos 50° \approx 95.3, \text{ which yields } c \approx 9.8.$$

EXAMPLE

▶ Find *B* as shown in the figure.

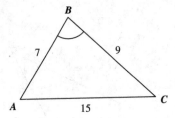

▶ Solve $b^2 = a^2 + c^2 - 2ac\cos B$ for $\cos B$. Then, substitute the information from the figure into the result to find the value of $\cos B$. Next, use the 2ND [cos⁻¹] calculator key to determine B.

$$b^2 = a^2 + c^2 - 2ac\cos B$$
$$\cos B = \frac{a^2 + c^2 - b^2}{2ac}$$
$$\cos B = \frac{9^2 + 7^2 - 15^2}{2(9)(7)}$$
$$\cos B \approx -0.7540, \text{ which yields}$$
$$B \approx \cos^{-1}(-0.7540) \approx 138.9°$$

EXAMPLE

A **resultant force** is the single force obtained when two or more forces act at a point concurrently. Two forces of 30 pounds and 45 pounds act on an object with an angle of 60° between them. Find the magnitude of the resultant force to the nearest pound.

Let $\overrightarrow{AC}$ represent the 30-pound force and $\overrightarrow{CB}$ represent the 45-pound force. Sketch a parallelogram containing $\overrightarrow{AC}$ and $\overrightarrow{CB}$ placed tip to end. Draw $\overrightarrow{AB}$ as the resultant force that joins the beginning to the end in a straight line as shown below.

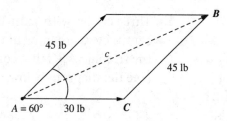

Because A and C are consecutive angles of a parallelogram, they are supplementary. Thus, $C = 120°$. Use the Law of Cosines to find the magnitude, c, of the resultant force $\overrightarrow{AB}$.

$$c^2 = 30^2 + 45^2 - 2(30)(45)\cos(120°)$$
$$c^2 = 900 + 2{,}025 - 2{,}700\cos(120°)$$
$$c^2 = 4{,}275$$
$$c = \sqrt{4{,}275} \approx 65.4$$

Therefore, to the nearest pound, the magnitude of the resultant force is 65 pounds.

EXERCISE 5-1

Solve as indicated. (Round answers to one decimal place, as needed.)

1. Solve for b.

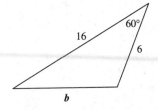

2. Solve for θ.

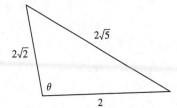

3. Solve for *a*.

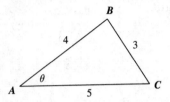

4. Solve for *θ*.

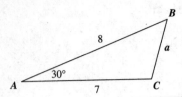

5. Find to the nearest tenth the magnitude of the resultant force when forces of 2.6 and 4.3 pounds act on an object with an angle of 40° between them.

6. A vertical tower 30 feet high is on a hill that forms an angle of 15° with the horizontal. How long is a guy wire that runs from the top of the tower to a point 20 feet from the base of the tower? (See the diagram below.)

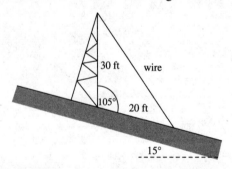

7. Find the measure of the smallest angle of the triangle whose sides are 4.3, 5.1, and 6.3.

8. Solve triangle *ABC* given *a* = 9, *b* = 7, and *c* = 5.

9. Solve triangle *ABC* given *a* = 3.2, *b* = 2.2, and *C* = 75.3°.

10. Solve triangle *ABC* given *B* = 110°, *a* = 25, and c = 15.

11. A ship is supposed to travel directly from city *A* to city *B*. The distance between the cities is 20 kilometers. After traveling a distance of 8 kilometers, the captain discovers that the ship has been traveling 18° off course. At this point, how far is the ship from city *B*?

12. The lengths of two adjacent sides of a parallelogram are 6 centimeters and 8 centimeters. The included angle is 67°. Find the length of the longest diagonal.

13. Three circles with radii 1, 3, and 4 centimeters are tangent to one another. Solve the triangle *ABC* that connects their centers. (See the diagram below.)

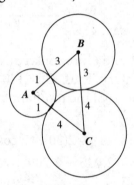

14. Two planes leave an airfield at the same time. One flies at an angle of 30° east of due north at 250 kilometers per hour, and the other flies at 45° east of due south at 300 kilometers per hour. How far apart are the planes after 2 hours? (See the diagram below.)

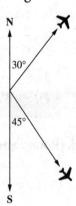

15. Points *B* and *C* are on opposite ends of a lake. If point *A* is 1,500 feet from point *B* and 2,000 feet from point *C* and the angle at *A* is 50°, how long is the lake?

16. Find the measure of the largest angle of a triangle whose side lengths are 2.9, 3.3, and 4.1.

17. True or False? The three sides $a = 12$, $b = 22$, and $c = 13$ determine a unique triangle.

18. True or False? The three angles $A = 12°$, $B = 82°$, and $A = 84°$ determine a unique triangle.

19. True or False? The three angles $A = 14°$, $B = 82°$, and $A = 84°$ determine a unique triangle.

20. True or False? The Law of Cosines can be used to solve a triangle given SAS.

Law of Sines (ASA or AAS)

The **Law of Sines** consists of equations for solving oblique triangles when you are given two angles and the included side (abbreviated ASA) or two angles and a non-included side (abbreviated AAS) of a triangle.

▶ **The Law of Sines**

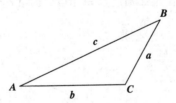

For the triangle shown, the Law of Sines is as follows:

$$\frac{\sin A}{a} = \frac{\sin B}{b} = \frac{\sin C}{c}$$

Keep in mind that the Law of Sines also applies to right triangles.

▶ Given triangle *ABC*, with $B = 36°$, $C = 104°$, and $b = 12$, find c.

▶ Draw a sketch.

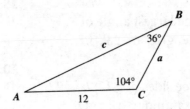

▶ Substitute the information from the figure into $\dfrac{\sin B}{b} = \dfrac{\sin C}{c}$, and then solve for c.

$$\frac{\sin 36°}{12} = \frac{\sin 104°}{c}$$

$$c = \frac{12 \sin 104°}{\sin 36°} \approx 19.8$$

▶ Given triangle *ABC* with $A = 40°$, $C = 104°$, and $b = 12$, solve for a.

▶ Draw a sketch.

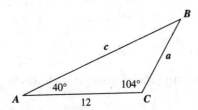

▶ Because the sum of the angles of a triangle equals 180°, it follows that

$B = 180° - (40° + 104°) = 36°$. Substitute the information from the

figure into $\dfrac{\sin B}{b} = \dfrac{\sin A}{a}$, and then solve for a.

$$\frac{\sin 36°}{12} = \frac{\sin 40°}{a}$$

$$a = \frac{12 \sin 40°}{\sin 36°} \approx 13.1$$

EXAMPLE

▶ Two forces applied on a heavy carton make angles of 35° and 48° with the resultant force of 150 pounds. Find to the nearest pound the magnitude of the larger force.

▶ Draw a sketch to represent the given forces with $\overrightarrow{AC}$ representing the larger force with magnitude, V, and $\overrightarrow{AB}$ as the resultant force of 150 pounds.

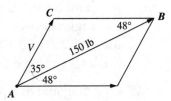

▶ Alternate interior angles are congruent, and the sum of the angles of a triangle is 180°. Hence, $C = 180° - (35° + 48°) = 97°$. Use the Law of Sines to find V because it is the magnitude of the larger force (given that it is opposite the 48° angle, making it larger than the side opposite the 35° angle in triangle ABC).

$$\frac{V}{\sin 48°} = \frac{150}{\sin 97°}$$

▶ Solving for V yields $V = \dfrac{150 \sin(48°)}{\sin(97°)} \approx 112.3$. Thus, to the nearest pound, the magnitude of the larger force is 112 pounds.

EXERCISE 5-2

Solve for the indicated unknown(s). (Round answers to one decimal place, as needed.)

1.

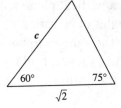

2.

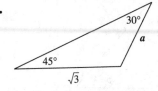

3.

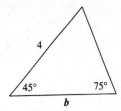

4.

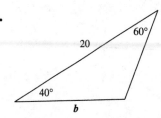

5.

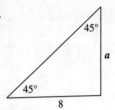

6.

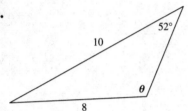

7.

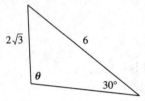

8.

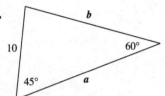

9.

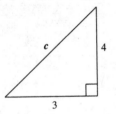

10.

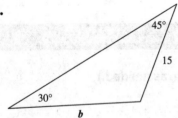

11.

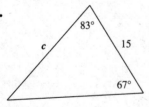

12.

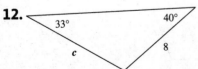

13.

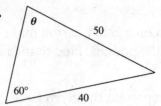

14. Two forces act on an object so that the resultant force is 50 pounds. The measure of the angles between the resultant and the two forces is 25° and 42°. Find the magnitude of the larger applied force to the nearest pound.

15. One diagonal of a parallelogram is 20 centimeters long and at one end makes angles of 20° and 40° with the sides of the parallelogram. Find the length of the parallelogram's sides.

16. From two points on shore, the angles A and B from the ground to a light at C are 16° and 58°, respectively. If AB is 1,500 feet, find the distance, d, from the shore. (See the diagram below.)

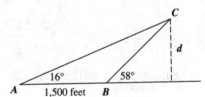

17. Two observers at points A and B are on the same side of a river and are 2,500 feet apart. They both spot an elk at point C on the opposite shore. The angle BAC is 78.5° and the angle ABC is 47.3°. Find the distance from B to C.

18. Two women 400 feet apart observe a kite between them that is in a vertical plane with them. The respective angles of elevation of the kite are observed by the women to be 73° and 50.4°. In the figure shown, find the height of the kite above the ground.

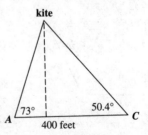

19. A weight is suspended from a rope strung between two vertical poles as shown in the figure. How far is the weight from the left pole?

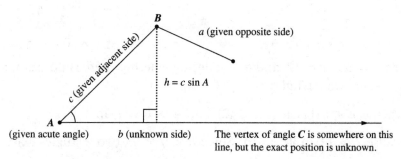

25 feet

20. True or False? A unique triangle can be constructed given three angles.

Law of Sines Ambiguous Case (SSA)

When you are solving oblique triangles, the ambiguous case arises when the only specifications given for a proposed triangle are the lengths of two sides and a non-included angle of the triangle (abbreviated SSA). In this circumstance, three situations can occur: (1) no triangle exists, (2) one unique triangle exists, or (3) two distinct triangles may satisfy the conditions given.

Consider a proposed triangle in which you are given a, c, and A. If A is obtuse and $a > c$, then there exists one unique triangle; but if $a \leq c$, no such triangle exists. If A is a right angle and $a > c$, then there exists one unique right triangle; but if $a \leq c$, no such triangle exists.

If A is acute, then there are three possibilities: no triangle, one unique triangle, or two distinct triangles. Here is a visual depiction of the dilemma that occurs (where $h = c \sin A$ is the distance from angle B's vertex to the opposite side):

> In mathematics, a *unique* triangle is the "one and only one."

B

a (given opposite side)

c (given adjacent side)

$h = c \sin A$

A
(given acute angle)

b (unknown side)

The vertex of angle C is somewhere on this line, but the exact position is unknown.

As a result, the following outcomes are possible (illustrated below):

▶ If $a \geq c$, a unique triangle exists.

▶ If $a = h$, one unique right triangle exists.

▶ If $a < h$, no such triangle exists.

▶ If $h < a < c$, two distinct triangles exist.

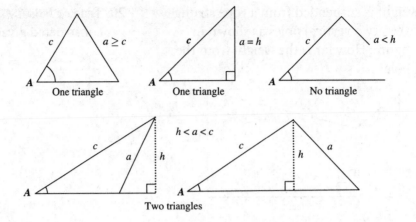

Given $A = 38°$, $c = 12$, and $a = 7$, decide whether there is no triangle, one triangle, or two triangles.

A is acute, $a < c$ (because $7 < 12$), and $a < h = c\sin A$

(because $7 < 12\sin 38° \approx 7.39$). Since $7 < h$, no triangle exists.

Given $A = 42°$, $c = 12$, and $a = 22$, decide whether there is no triangle, one triangle, or two triangles.

A is acute and $a > c$ (because $22 > 12$). Thus, there is one triangle.

Given $A = 30°$, $c = 12$, and $a = 7$, decide whether there is no triangle, one triangle, or two triangles.

A is acute, $a < c$ (because $7 < 12$), and $a > h = c\sin A$

(because $7 > 12\sin 30° = 6$). Since $h < 7 < c$, two triangles exist.

EXERCISE 5-3

For the given information, decide whether there is no triangle, one triangle, or two triangles.

1. Given $A = 120°$, $c = 12$, and $a = 16$

2. Given $A = 38°$, $c = 12$, and $a = 22$

3. Given $A = 45°$, $c = 12$, and $a = 7$

4. Given $A = 104°$, $c = 38$, and $a = 22$

5. Given $A = 15°$, $c = 8$, and $a = 5$

6. Given $A = 45°$, $c = 38$, and $a = 42$

7. Given $A = 74°$, $c = 22$, and $a = 22$

8. Given $A = 60°$, $c = 38$, and $a = 16$

9. Given $A = 65°$, $c = 38$, and $a = 22$

10. Given $A = 30°$, $c = 44$, and $a = 22$

11. Given $A = 30°$, $c = 66$, and $a = 44$

12. Given $A = 45°$, $c = 16$, and $a = 16$

13. Given $A = 100°$, $c = 122$, and $a = 145$

14. True or False? A unique triangle is determined when you are given ASA.

15. True or False? A unique triangle is determined when you are given SSA.

Solving General Triangles

The Law of Cosines and the Law of Sines apply to both right triangles and oblique triangles. The following table presents a situation; the number of triangles that result; and whether the Law of Cosines, the Law of Sines, or neither is most appropriate for the situation.

> Keep in mind that for a triangle to exist, the information given must neither violate the triangle inequality nor the constraint that a triangle's interior angles sum to 180°.

Information Given in the Problem	Number of Triangles	Use
Three sides (SSS), and the sum of the lengths of the two smaller sides is greater than the length of the larger side.	One triangle	Law of Cosines
Three sides (SSS), and the sum of the lengths of the two smaller sides is less than or equal to the length of the larger side.	No triangle	Neither
Two sides and the included angle (SAS).	One triangle	Law of Cosines
Two angles and the included side (ASA), and the sum of the given angles is less than 180°.	One triangle	Law of Sines
Two angles and a non-included side (AAS), and the sum of the given angles is less than 180°.	One triangle	Law of Sines

Information Given in the Problem	Number of Triangles	Use
Two angles and either the included side (ASA) or a non-included side (AAS), and the sum of the given angles is greater than or equal to 180°.	No triangle	Neither
Two sides and a non-included obtuse angle (SSA), and the length of the side opposite the given angle is greater than the length of the side adjacent to the given angle.	One triangle	Law of Sines
Two sides and a non-included obtuse angle (SSA), and the length of the side opposite the given angle is less than or equal to the length of the side adjacent to the given angle.	No triangle	Neither
Two sides and a non-included acute angle (SSA), and the length of the side opposite the given angle is greater than or equal to the length of the side adjacent to the given angle.	One triangle	Law of Sines
Two sides and a non-included acute angle (SSA), and the length of one side falls between the length of the altitude from the vertex where the two given sides meet and the length of the other side.	Two triangles	Law of Sines
Two sides and a non-included acute angle (SSA), and the length of the altitude from the vertex where the two given sides meet falls between the lengths of the two given sides.	No triangle	Neither
Three angles (AAA).	No unique triangle	Neither

Note to the reader: When looking at the examples and working through the questions that follow, realize there can be multiple ways to find the solution in a given situation. You might think of ways to reach the correct solutions other than the ones shown.

EXAMPLE

Solve triangle ABC, given $A = 40°$, $a = 50$, and $b = 30$.

You are given two sides and a non-included acute angle (SSA), and the length of the side opposite the given angle is greater than the length of the side adjacent to the given angle, so there is one triangle.

Using the Law of Sines yields $\dfrac{30}{\sin B} = \dfrac{50}{\sin 40°}$. Hence,

$\sin B = \dfrac{30 \sin 40°}{50} \approx 0.3857$. Given $\sin^{-1}(0.3857) \approx 22.7°$, then either $B \approx 22.7°$ or $B \approx 180° - 22.7° = 157.3°$. The latter value will not work because $40° + 157.3° > 180°$. It follows that $B \approx 22.7°$, and $C \approx 180° - 40° - 22.7° = 117.3°$. Using this result and again applying the Law of Sines gives $\dfrac{50}{\sin 40°} = \dfrac{c}{\sin 117.3°}$. Hence, $c = \dfrac{50 \sin 117.3°}{\sin 40°} \approx 69.1$.

Therefore, the triangle is solved.

EXAMPLE

Solve triangle ABC, given $a = 15$, $b = 25$, and $c = 28$.

You are given three sides (SSS), and the sum of the lengths of the two smaller sides is greater than the length of the larger side, so there is one triangle. Applying the Law of Cosines, solve for A.

$\cos A = \dfrac{25^2 + 28^2 - 15^2}{2(25)(28)} \approx 0.8457$

$A \approx 32.3°$

Applying the Law of Cosines a second time, solve for B.

$\cos B = \dfrac{15^2 + 28^2 - 25^2}{2(15)(28)} \approx 0.4571$

$B \approx 62.8°$

Then $C \approx 180° - 32.3° - 62.8° = 84.9°$. Therefore, the triangle is solved.

▶ Solve triangle ABC, given $A = 40°$, $B = 140°$, and $b = 30$.

You are given two angles and the non-included side (AAS), and the sum of the given angles 40° and 140° equals 180°. So, there is no triangle.

EXERCISE 5-4

For questions 1 to 6, solve triangle *ABC* using the provided information. (Round answers to one decimal place, as needed.)

1. $A = 35°$, $C = 90°$, and $c = 112$

2. $a = 2$, $b = 8$, and $c = 10$

3. $a = 4$, $b = 9$, and $c = 6$

4. $B = 70°$, $C = 25°$, and $c = 4$

5. $A = 62°$, $B = 120°$, and $a = 15$

6. $a = 520$, $c = 422$, and $A = 130°$

For questions 7 to 20, solve triangle *ABC* for the indicated part. (Round answers to one decimal place, as needed.)

7. $a = 7.5$, $b = 4.6$, and $c = 7$. Find B.

8. $B = 48°$, $C = 90°$, and $c = 220$. Find b.

9. $A = 23°$, $C = 90°$, and $c = 345$. Find a.

10. $a = 15$, $A = 25°$, and $B = 40°$. Find b.

11. $a = 315$, $b = 460$, and $A = 42°$. Find B.

12. $a = 17$, $c = 14$, and $B = 30°$. Find b.

13. $a = 25$, $b = 38$, and $C = 11°$. Find c.

14. $B = 54°$, $C = 90°$, and $c = 180$. Find a.

15. $A = 32°$, $C = 90°$, and $a = 75$. Find c.

16. $A = 58°$, $C = 90°$, and $b = 38$. Find c.

17. $c = 190$, $a = 150$, and $C = 85.2°$. Find A.

18. $b = 17$, $a = 12$, and $A = 25°$. Find B.

19. $A = 24.2°$, $B = 56.5°$, and $a = 32$. Find b.

20. $c = 0.5$, $b = 0.8$, and $A = 70°$. Find a.

For questions 21 to 30, solve as indicated. (Round answers to one decimal place, as needed.)

21. A hot-air balloon is anchored to the ground at point A by a rope that is 150 yards long and at point B by a rope that is 120 yards long. If the angle between the two ropes is 85°, what is the distance between points A and B? (See the diagram below.)

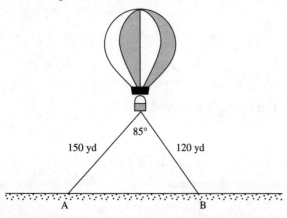

22. A fence is built around a triangular garden. Two sides of the fence measure 120 feet and 80 feet, and the angle between the two sides is 45°. Find the length of the other side.

23. A flagpole of height 110 feet sits on top of a hill. From a point A on level ground away from the base of the hill, the angles of elevation to the top B and bottom C of the flagpole are measured as 48.50° and 39.75°, respectively. Determine the height of the hill. (See the diagram on the following page.)

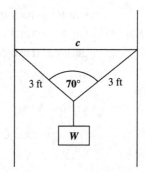

110 ft

48.50°

39.75°

24. The base of an isosceles triangle is 30 and the base angles are 50°. Find the equal sides and the altitude of the triangle.

25. A survey line must cross a swampy area. At point *A* the surveyor sighted a point *B* at an angle of 52° at a distance of 1,550 feet from point *A*. At point *B* the surveyor turned an angle of 90° and ran a line *BC*. If *C* is on the line through the swamp, how far is point *C* from point *B*? (See the diagram below.)

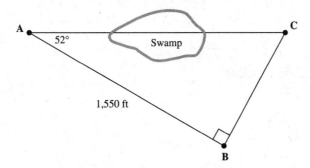

26. Two photographers each take a picture of the same elk drinking at a river's edge. As shown in the diagram below, the two photographers are 400 feet apart, one at point *A* and one at point *B*. Find the distance from the photographer at point *A* to the elk.

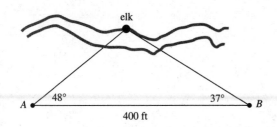

27. A brace, *c*, is to be installed between two posts that are supporting a weight, *W*, held by two wires. The lengths of the wires and the angle between them are as shown in the figure below. Find the length of the brace, *c*.

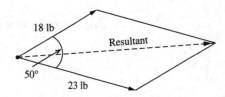

28. Two forces of 18 and 23 pounds act on an object. If the angle between the two forces is 50°, find the magnitude of their resultant. (See the diagram below.)

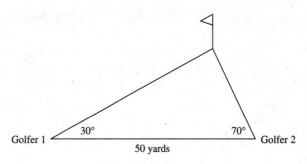

29. The angle of elevation from an observer to a flagpole is 21°. If the observer approaches the flagpole by 24 meters, the angle of elevation becomes 35°. Determine the height of the flagpole.

30. Two golfers are aiming for the green. The golfers are 50 yards apart, and the angles are as shown in the diagram below. How far is golfer 1 from the flagstick?

Area of a General Triangle Using Trigonometry

The familiar formula for the area of a triangle is $\text{Area} = \frac{1}{2}bh$, where b is the length of one side and h is the height of the triangle drawn to that side (or an extension of it). Using trigonometry, you can determine that for a general triangle ABC, with given sides a and b and the included angle C, $h = a\sin C$ (illustrated below). Substituting into the formula $\text{Area} = \frac{1}{2}bh$ yields $\text{Area} = \frac{1}{2}b \cdot a\sin C = \frac{1}{2}ab\sin C$, where a and b are any two sides of a triangle and C is the included angle.

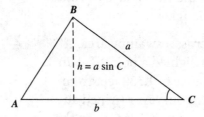

This formula works for all triangles.

Find the area of the triangle shown below.

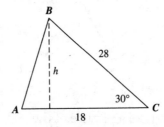

$\text{Area} = \frac{1}{2}(28)(18)\sin 30° = 126.$

EXAMPLE

Find the area of the triangle shown below.

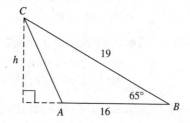

$$\text{Area} = \frac{1}{2}(19)(16)\sin 65° \approx 137.8$$

EXERCISE 5-5

For questions 1 to 11, determine the area of triangle *ABC* using the provided information. (Round answers to one decimal place, as needed).

1. $a = 25$, $b = 30$, and $C = 60°$.

2. $a = 20$, $b = 15$, and $C = 150°$.

3. $a = 18$, $b = 10$, and $C = 120°$.

4. $a = 14$, $c = 17$, and $B = 50°$.

5. $a = 24$, $b = 16$, and $C = 90°$.

6.

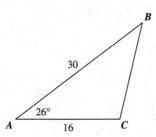

7.

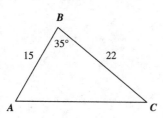

8.

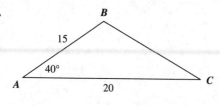

9.

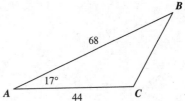

10.

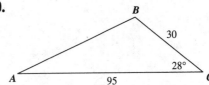

11.

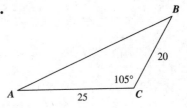

For questions 12 to 15, solve as indicated.

12. Find the area of the given triangle.

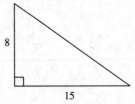

13. A jib sail is in the form of a triangle. One side is 8 feet long and another side is 6 feet long, and the angle included between the sides is 54°. How much material in feet² is contained in the sail?

14. A floor tile is made in the form of an equilateral triangle whose sides measure 10 inches each. What is the area of the tile?

15. A quilt pattern is made in the shape of a regular hexagon whose sides measure 3 inches each. What is the area of the pattern?

CHAPTER 6

Trigonometric Functions of Any Angle

Definitions of the Trigonometric Functions

Let θ be any angle in standard position and (x, y) be a point on θ's terminal side such that the line segment, $r = \sqrt{x^2 + y^2}$, from the origin to (x, y) is not zero (illustrated below).

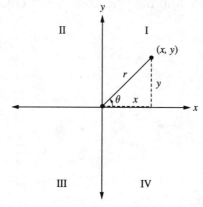

The trigonometric functions of θ are defined as follows:
Definitions of the Trigonometric Functions of Any Angle θ

$$\sin\theta = \frac{y}{r} \quad \cos\theta = \frac{x}{r}$$

$$\tan\theta = \frac{y}{x},\ x \neq 0 \quad \cot\theta = \frac{x}{y},\ y \neq 0$$

$$\sec\theta = \frac{r}{x},\ x \neq 0 \quad \csc\theta = \frac{r}{y},\ y \neq 0$$

Because $r = \sqrt{x^2 + y^2} \neq 0$, the sine and cosine functions are defined for all real values of θ. The tangent and secant are undefined when $x = 0$; similarly, the cotangent and cosecant are undefined when $y = 0$.

EXAMPLE

Using the figure shown, find the sine, cosine, and tangent of θ.

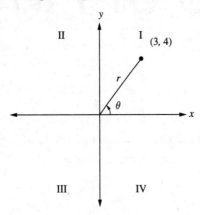

From the figure $x = 3$, $y = 4$, and $r = \sqrt{x^2 + y^2} = \sqrt{3^2 + 4^2} = \sqrt{9 + 16} = \sqrt{25} = 5$

Thus, $\sin\theta = \frac{y}{r} = \frac{4}{5}$, $\cos\theta = \frac{x}{r} = \frac{3}{5}$, and $\tan\theta = \frac{y}{x} = \frac{4}{3}$.

EXAMPLE

Using the figure shown, find the sine, cosine, and tangent of θ.

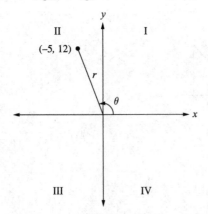

From the figure $x = -5$, $y = 12$, and $r = \sqrt{x^2 + y^2} = \sqrt{(-5)^2 + 12^2}$

$= \sqrt{25 + 144} = \sqrt{169} = 13$.

Thus, $\sin\theta = \frac{y}{r} = \frac{12}{13}$, $\cos\theta = \frac{x}{r} = \frac{-5}{13} = -\frac{5}{13}$, and $\tan\theta = \frac{y}{x} = \frac{12}{-5} = -\frac{12}{5}$.

The signs of the sine, cosine, and tangent trigonometric functions in the four quadrants follow the pattern shown below.

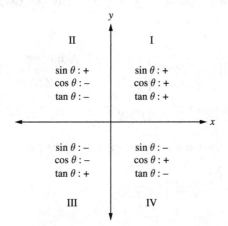

The signs of cosecant, secant, and cotangent have the same sign pattern as their corresponding reciprocals, so it is sufficient to memorize the signs of sine, cosine, and tangent only.

EXAMPLE

The point $(-2, -2)$ lies on the terminal side of θ. Find the sine, cosine, and tangent of θ. (Write the exact answer in simplest radical form for irrational answers.)

From the given information $x = -2$, $y = -2$, and $r = \sqrt{x^2 + y^2}$
$= \sqrt{(-2)^2 + (-2)^2} = \sqrt{4 + 4} = \sqrt{8} = 2\sqrt{2}$.

Thus, $\sin\theta = \dfrac{y}{r} = \dfrac{-2}{2\sqrt{2}} = -\dfrac{1}{\sqrt{2}} = -\dfrac{\sqrt{2}}{2}$, $\cos\theta = \dfrac{x}{r} = \dfrac{-2}{2\sqrt{2}} = -\dfrac{1}{\sqrt{2}}$

$= -\dfrac{\sqrt{2}}{2}$, and $\tan\theta = \dfrac{y}{x} = \dfrac{-2}{-2} = 1$.

EXAMPLE

The point $(3, -4)$ lies on the terminal side of θ. Find the secant, cosecant, and cotangent of θ. (Write the exact answer in simplest radical form for irrational answers.)

From the given information $x = 3$, $y = -4$, and $r = \sqrt{x^2 + y^2}$
$= \sqrt{(3)^2 + (-4)^2} = \sqrt{9 + 16} = \sqrt{25} = 5$. Thus, $\sec\theta = \dfrac{r}{x} = \dfrac{5}{3}$,

$\csc\theta = \dfrac{r}{y} = \dfrac{5}{-4} = -\dfrac{5}{4}$, and $\cot\theta = \dfrac{x}{y} = \dfrac{3}{-4} = -\dfrac{3}{4}$.

The values of the trigonometric functions of an angle θ are the same, regardless of the choice of the point (x, y) selected on the terminal side of θ.

EXAMPLE

> Given θ is in quadrant II and $\cos\theta = -\dfrac{4}{5}$, find the exact values of $\sin\theta$ and $\tan\theta$.

> As θ is in quadrant II and $\cos\theta = -\dfrac{4}{5} = \dfrac{\text{Adjacent}}{\text{Hypotenuse}}$, you can let $r = 5$ and $x = -4$, from which you have $y = \sqrt{5^2 - (-4)^2} = \sqrt{25 - 16} = \sqrt{9} = 3$ (which is positive because θ is in quadrant II). So, $\sin\theta = \dfrac{y}{r} = \dfrac{3}{5}$ and $\tan\theta = \dfrac{y}{x} = \dfrac{3}{-4} = -\dfrac{3}{4}$.

EXERCISE 6-1

For questions 1 to 10, solve as indicated. (Write the exact answer in simplest radical form for irrational answers.)

1. The point $(2\sqrt{6}, -1)$ lies on the terminal side of θ. Find the sine, cosine, and tangent of θ.

2. The point $(0.6, -0.8)$ lies on the terminal side of θ. Find the sine, cosine, and tangent of θ.

3. The point $(-3, \sqrt{7})$ lies on the terminal side of θ. Find the sine, cosine, and tangent of θ.

4. The point $(-8, -15)$ lies on the terminal side of θ. Find the sine, cosine, and tangent of θ.

5. The point $(10\sqrt{3}, 10)$ lies on the terminal side of θ. Find the sine, cosine, and tangent of θ.

6. The point $(-3, \sqrt{7})$ lies on the terminal side of θ. Find the secant, cosecant, and cotangent of θ.

7. The point $(-9, -40)$ lies on the terminal side of θ. Find the secant, cosecant, and cotangent of θ.

8. The point $(0.8, -0.6)$ lies on the terminal side of θ. Find the secant, cosecant, and cotangent of θ.

9. The point $(-12, 5)$ lies on the terminal side of θ. Find the secant, cosecant, and cotangent of θ.

10. The point $(7, 24)$ lies on the terminal side of θ. Find the secant, cosecant, and cotangent of θ.

For questions 11 to 15, for the given information, name the quadrants in which θ could lie.

11. $\cos\theta > 0$

12. $\tan\theta < 0$

13. $\csc\theta > 0$

14. $\cot\theta > 0$

15. $\sin\theta < 0$

For questions 16 to 20, for the given information, name the quadrant in which θ lies.

16. $\tan\theta > 0$ and $\cos\theta < 0$

17. $\sin\theta > 0$ and $\sec\theta < 0$

18. $\csc\theta > 0$ and $\cos\theta > 0$

19. $\sec\theta < 0$ and $\cot\theta > 0$

20. $\sin\theta > 0$ and $\cot\theta < 0$

For questions 21 to 25, solve as indicated. (Write the exact answer in simplest radical form for irrational answers.)

21. Given θ is in quadrant III and $\cos\theta = -\dfrac{5}{13}$, find the exact values of $\sin\theta$ and $\tan\theta$.

22. Given θ is in quadrant IV and $\sin\theta = -\dfrac{6}{10}$, find the exact values of $\cos\theta$ and $\tan\theta$.

23. Given θ is in quadrant I and $\sec\theta = \dfrac{41}{9}$, find the exact values of $\sin\theta$ and $\cos\theta$.

24. Given θ is in quadrant II and $\tan\theta = -\dfrac{7}{24}$, find the exact values of $\sin\theta$ and $\cos\theta$.

25. Given θ is in quadrant III and $\sin\theta = -\dfrac{15}{17}$, find the exact values of $\cos\theta$ and $\tan\theta$.

Trigonometric Functions of Complementary Angles

In right triangle ABC shown below, with $C = 90°$, the acute angles α and β are complementary.

> Two acute angles are complementary if their sum is 90° $\left(\text{or } \dfrac{\pi}{2}\right)$.

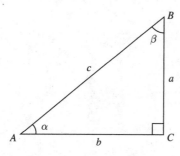

From the figure, you have the following **cofunction** relationships:

Cofunction Relationships of Complementary Angles

$$\sin\beta = \cos(90° - \beta) = \cos\left(\frac{\pi}{2} - \beta\right) = \frac{b}{c} = \cos\alpha$$

$$\cos\beta = \sin(90° - \beta) = \sin\left(\frac{\pi}{2} - \beta\right) = \frac{a}{c} = \sin\alpha$$

$$\tan\beta = \cot(90° - \beta) = \cot\left(\frac{\pi}{2} - \beta\right) = \frac{b}{a} = \cot\alpha$$

$$\sec\beta = \csc(90° - \beta) = \csc\left(\frac{\pi}{2} - \beta\right) = \frac{c}{a} = \csc\alpha$$

$$\csc\beta = \sec(90° - \beta) = \sec\left(\frac{\pi}{2} - \beta\right) = \frac{c}{b} = \sec\alpha$$

$$\cot\beta = \tan(90° - \beta) = \tan\left(\frac{\pi}{2} - \beta\right) = \frac{a}{b} = \tan\alpha$$

These relationships occur in pairs and are valid for any pair of complementary angles. Each function of a pair is the cofunction of the other: Sine and cosine are cofunctions of each other; secant and cosecant are cofunctions of each other; and tangent and cotangent are cofunctions of each other. In general, any function of an acute angle equals the corresponding cofunction of its complementary angle.

EXAMPLE

$\sin 55° = \cos 35°$

$\tan \dfrac{\pi}{6} = \cot \dfrac{\pi}{3}$

If $\sin 40° = 0.6428$, then $\cos 50° = 0.6428$.

Knowing the cofunction relationships can be helpful in solving certain trigonometric equations.

EXAMPLE

Solve $\sin(2\theta + 5°) = \cos(33°)$.

$2\theta + 5°$ and $33°$ are complementary angles.

Therefore,

$$2\theta + 5° = 90° - 33°$$
$$2\theta + 5° = 57°$$
$$2\theta = 52°$$
$$\theta = 26°$$

EXERCISE 6-2

For questions 1 to 10, select the best answer choice.

1. Which of the following equals sin 32°?

   a. cos 32°

   b. cos 58°

   c. sin 58°

   d. csc 32°

2. Which of the following equals cos 15°?

   a. sin 75°

   b. sec 15°

   c. csc 75°

   d. sin 15°

3. Which of the following equals sec 25°?

   a. csc 25°

   b. cos 65°

   c. csc 65°

   d. sec 65°

4. Which of the following equals $\csc\dfrac{13\pi}{30}$?

   a. $\sin\dfrac{13\pi}{30}$

   b. $\csc\dfrac{\pi}{15}$

   c. $\sec\dfrac{13\pi}{30}$

   d. $\sec\dfrac{\pi}{15}$

5. Which of the following equals $\cos\dfrac{\pi}{5}$?

   a. $\sin\dfrac{4\pi}{5}$

   b. $\cos\dfrac{4\pi}{5}$

   c. $\sin\dfrac{3\pi}{10}$

   d. $\cos\dfrac{3\pi}{10}$

6. If $\tan\alpha = \cot\beta$, then which of the following must be true?

   a. $\alpha + \beta = \pi$

   b. $\alpha - \beta = \dfrac{\pi}{2}$

   c. $\alpha + \beta = \dfrac{\pi}{2}$

   d. $\alpha - \beta = \pi$

7. If sin 35° = 0.5736, then which of the following must be true?

   a. sec 35° = 0.5736

   b. cos 55° = 0.5736

   c. sin 55° = 0.5736

   d. csc 55° = 0.5736

8. Given the triangle shown, which of the following equals the ratio $\dfrac{5}{13}$?

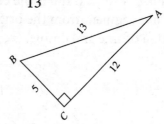

   a. sin *B*

   b. cos *A*

   c. sin (90° − *A*)

   d. cos (90° − *A*)

9. Given the triangle shown, which of the following is NOT equal to the ratio $\dfrac{15}{17}$?

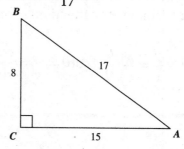

   a. sin *B*

   b. cos *A*

   c. sin *A*

   d. $\cos\left(\dfrac{\pi}{2} - B\right)$

For questions 10 to 20, solve for θ.

10. sin 28° = cos θ

11. cos 54° = sin θ

12. $\tan\dfrac{\pi}{3} = \cot\theta$

13. sin 48° = cos (θ + 7°)

14. tan 63° = cot (θ − 40°)

15. sin (2θ + 5°) = cos (15°)

16. $\sin\left(\dfrac{1}{2}\theta\right) = \cos\left(\dfrac{5}{2}\theta + \dfrac{\pi}{6}\right)$

17. $\sec\left(5\theta + \dfrac{\pi}{12}\right) = \csc\left(3\theta - \dfrac{\pi}{4}\right)$

18. $\cot 2\theta = \tan\theta$

19. $\sin(7\theta + 15°) = \cos(3\theta + 40°)$

20. $\csc\left(\dfrac{1}{3}\theta + 20°\right) = \sec 50°$

The Unit Circle

The **unit circle** is the circle centered at the origin with radius 1. It has the equation $x^2 + y^2 = 1$. If θ is the central angle in standard position that is formed by the line segment from the origin to a point (x, y) on the unit circle, then, by definition of sine and cosine, $(x, y) = (\cos\theta, \sin\theta)$. (See illustration below.)

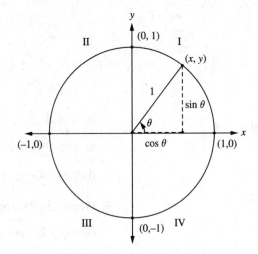

From the unit circle, given that $x = \cos\theta$ and $y = \sin\theta$, you can obtain the other trigonometric functions for θ as follows:

$$\tan\theta = \frac{y}{x} = \frac{\sin\theta}{\cos\theta}, \quad \cot\theta = \frac{x}{y} = \frac{\cos\theta}{\sin\theta}, \quad \sec\theta = \frac{1}{x} = \frac{1}{\cos\theta} \quad \text{and} \quad \csc\theta = \frac{1}{y} = \frac{1}{\sin\theta}$$

EXAMPLE

▶ The point $\left(-\dfrac{\sqrt{3}}{2}, -\dfrac{1}{2}\right)$ is a point on the unit circle corresponding to an angle θ in standard position. Find the sine, cosine, and tangent of θ. (Write the exact answer in simplest radical form for irrational answers.)

▶ From the given information, $\sin\theta = -\dfrac{1}{2}$, $\cos\theta = -\dfrac{\sqrt{3}}{2}$,

and $\tan\theta = \dfrac{\sin\theta}{\cos\theta} = \dfrac{-\dfrac{\sqrt{3}}{2}}{-\dfrac{1}{2}} = \sqrt{3}$.

EXAMPLE

The point $\left(\dfrac{4}{5}, -\dfrac{3}{5}\right)$ is a point on the unit circle corresponding to an angle θ in standard position. Find the secant, cosecant, and cotangent of θ. (Write the exact answer in simplest radical form for irrational answers.)

From the given information, $\sin\theta = -\dfrac{3}{5}$ and $\cos\theta = \dfrac{4}{5}$.

Then, $\sec\theta = \dfrac{1}{\cos\theta} = \dfrac{1}{\left(\dfrac{4}{5}\right)} = \dfrac{5}{4}$, $\csc\theta = \dfrac{1}{\sin\theta} = \dfrac{1}{\left(-\dfrac{3}{5}\right)} = -\dfrac{5}{3}$, and

$\cot\theta = \dfrac{\cos\theta}{\sin\theta} = \dfrac{\left(\dfrac{4}{5}\right)}{\left(-\dfrac{3}{5}\right)} = -\dfrac{4}{3}$.

EXERCISE 6-3

Solve as indicated. (Write the exact answer in simplest radical form for irrational answers.)

1. The point $\left(\dfrac{1}{2}, \dfrac{\sqrt{3}}{2}\right)$ is a point on the unit circle corresponding to an angle θ in standard position. Find the sine, cosine, and tangent of θ.

2. The point $\left(\dfrac{\sqrt{3}}{2}, -\dfrac{1}{2}\right)$ is a point on the unit circle corresponding to an angle θ in standard position. Find the sine, cosine, and tangent of θ.

3. The point $\left(-\dfrac{\sqrt{2}}{2}, \dfrac{\sqrt{2}}{2}\right)$ is a point on the unit circle corresponding to an angle θ in standard position. Find the sine, cosine, and tangent of θ.

4. The point $\left(\dfrac{12}{13}, \dfrac{5}{13}\right)$ is a point on the unit circle corresponding to an angle θ in standard position. Find the sine, cosine, and tangent of θ.

5. The point $\left(-\dfrac{1}{2}, -\dfrac{\sqrt{3}}{2}\right)$ is a point on the unit circle corresponding to an angle θ in standard position. Find the sine, cosine, and tangent of θ.

6. The point $\left(-\dfrac{9}{41}, -\dfrac{40}{41}\right)$ is a point on the unit circle corresponding to an angle θ in standard position. Find the sine, cosine, and tangent of θ.

7. The point $\left(\dfrac{\sqrt{2}}{2}, -\dfrac{\sqrt{2}}{2}\right)$ is a point on the unit circle corresponding to an angle θ in standard position. Find the sine, cosine, and tangent of θ.

8. The point $\left(-\dfrac{12}{13}, \dfrac{5}{13}\right)$ is a point on the unit circle corresponding to an angle θ in standard position. Find the sine, cosine, and tangent of θ.

9. The point $\left(-\dfrac{7}{25}, \dfrac{24}{25}\right)$ is a point on the unit circle corresponding to an angle θ in standard position. Find the sine, cosine, and tangent of θ.

10. The point $\left(\dfrac{36}{85}, -\dfrac{77}{85}\right)$ is a point on the unit circle corresponding to an angle θ in standard position. Find the sine, cosine, and tangent of θ.

11. The point $\left(\dfrac{1}{2}, \dfrac{\sqrt{3}}{2}\right)$ is a point on the unit circle corresponding to an angle θ in standard position. Find the secant, cosecant, and cotangent of θ.

12. The point $\left(\dfrac{\sqrt{3}}{2}, -\dfrac{1}{2}\right)$ is a point on the unit circle corresponding to an angle θ in standard position. Find the secant, cosecant, and cotangent of θ.

13. The point $\left(-\dfrac{\sqrt{2}}{2}, \dfrac{\sqrt{2}}{2}\right)$ is a point on the unit circle corresponding to an angle θ in standard position. Find the secant, cosecant, and cotangent of θ.

14. The point $\left(\dfrac{12}{13}, \dfrac{5}{13}\right)$ is a point on the unit circle corresponding to an angle θ in standard position. Find the secant, cosecant, and cotangent of θ.

15. The point $\left(-\dfrac{1}{2}, -\dfrac{\sqrt{3}}{2}\right)$ is a point on the unit circle corresponding to an angle θ in standard position. Find the secant, cosecant, and cotangent of θ.

16. The point $\left(-\dfrac{9}{41}, -\dfrac{40}{41}\right)$ is a point on the unit circle corresponding to an angle θ in standard position. Find the secant, cosecant, and cotangent of θ.

17. The point $\left(\dfrac{\sqrt{2}}{2}, -\dfrac{\sqrt{2}}{2}\right)$ is a point on the unit circle corresponding to an angle θ in standard position. Find the secant, cosecant, and cotangent of θ.

18. The point $\left(-\dfrac{12}{13}, \dfrac{5}{13}\right)$ is a point on the unit circle corresponding to an angle θ in standard position. Find the secant, cosecant, and cotangent of θ.

19. The point $\left(-\dfrac{7}{25}, \dfrac{24}{25}\right)$ is a point on the unit circle corresponding to an angle θ in standard position. Find the secant, cosecant, and cotangent of θ.

20. The point $\left(\dfrac{36}{85}, -\dfrac{77}{85}\right)$ is a point on the unit circle corresponding to an angle θ in standard position. Find the secant, cosecant, and cotangent of θ.

Trigonometric Functions of Quadrantal Angles

A quadrantal angle is an angle in standard position whose terminal side lies on an axis. The angles 0°, 90°, 180°, and 270° and all the angles coterminal with them are quadrantal angles.

Some examples of quadrantal angles are:

0°, 90°, 180°, 270°, 360°, 450°, and so on

−90°, −180°, −270°, −360°, −450°, and so on

$0, \dfrac{\pi}{2}, \pi, \dfrac{3\pi}{2}, 2\pi, \dfrac{5\pi}{2}$, and so on

$-\dfrac{\pi}{2}, -\pi, -\dfrac{3\pi}{2}, -2\pi, -\dfrac{5\pi}{2}$, and so on

Using the unit circle and the definitions of the trigonometric functions, you can determine the trigonometric functions for quadrantal angles.

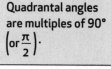

Quadrantal angles are multiples of 90° $\left(\text{or } \dfrac{\pi}{2}\right)$.

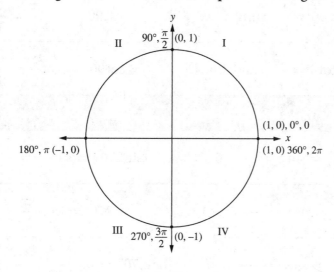

EXAMPLE

Given the point $(x, y) = (0, 1)$ on the unit circle corresponding to an angle of 90° in standard position, $\sin 90° = 1$, $\cos 90° = 0$,

$$\tan 90° = \frac{\sin 90°}{\cos 90°} = \frac{1}{0} = \text{undefined}, \; \sec 90° = \frac{1}{\cos 90°} = \frac{1}{0} = \text{undefined},$$

$$\csc 90° = \frac{1}{\sin 90°} = \frac{1}{1} = 1, \text{ and } \cot 90° = \frac{\cos 90°}{\sin 90°} = \frac{0}{1} = 0.$$

The use of the "=" sign to indicate that a trigonometric function is undefined does not have the customary meaning of "equals." This notation is used in math as a short way to describe a special situation in which an expression has no meaning.

The trigonometric values of the quadrantal angles from 0° (= 0 radians) to 360° (= 2π radians) are as follows.

$\theta = 0° = 0; (x, y) = (1, 0), r = 1$	$\theta = 90° = \dfrac{\pi}{2}; (x, y) = (0, 1), r = 1$
$\sin 0° = \sin 0 = \dfrac{0}{1} = 0$	$\sin 90° = \sin \dfrac{\pi}{2} = \dfrac{1}{1} = 1$
$\cos 0° = \cos 0 = \dfrac{1}{1} = 1$	$\cos 90° = \cos \dfrac{\pi}{2} = \dfrac{0}{1} = 0$
$\tan 0° = \tan 0 = \dfrac{0}{1} = 0$	$\tan 90° = \tan \dfrac{\pi}{2} = \dfrac{1}{0} = \text{undefined}$
$\sec 0° = \sec 0 = \dfrac{1}{1} = 1$	$\sec 90° = \sec \dfrac{\pi}{2} = \dfrac{1}{0} = \text{undefined}$
$\csc 0° = \csc 0 = \dfrac{1}{0} = \text{undefined}$	$\csc 90° = \csc \dfrac{\pi}{2} = \dfrac{1}{1} = 1$
$\cot 0° = \cot 0 = \dfrac{1}{0} = \text{undefined}$	$\cot 90° = \cot \dfrac{\pi}{2} = \dfrac{0}{1} = 0$
$\theta = 180° = \pi; (x, y) = (-1, 0), r = 1$	$\theta = 270° = \dfrac{3\pi}{2}; (x, y) = (0, -1), r = 1$
$\sin 180° = \sin \pi = \dfrac{0}{1} = 0$	$\sin 270° = \sin \dfrac{3\pi}{2} = \dfrac{-1}{1} = -1$
$\cos 180° = \cos \pi = \dfrac{-1}{1} = -1$	$\cos 270° = \cos \dfrac{3\pi}{2} = \dfrac{0}{1} = 0$
$\tan 180° = \tan \pi = \dfrac{0}{-1} = 0$	$\tan 270° = \tan \dfrac{3\pi}{2} = \dfrac{-1}{0} = \text{undefined}$
$\sec 180° = \sec \pi = \dfrac{1}{-1} = -1$	$\sec 270° = \sec \dfrac{3\pi}{2} = \dfrac{1}{0} = \text{undefined}$
$\csc 180° = \csc \pi = \dfrac{1}{0} = \text{undefined}$	$\csc 270° = \csc \dfrac{3\pi}{2} = \dfrac{1}{-1} = -1$
$\cot 180° = \cot \pi = \dfrac{-1}{0} = \text{undefined}$	$\cot 270° = \cot \dfrac{3\pi}{2} = \dfrac{0}{-1} = 0$

EXAMPLE

▶ Evaluate $\cos 0° - 4\sin 270° + 2\cot 90°$.

$\cos 0° - 4\sin 270° + 2\cot 90° = 1 - 4(-1) + 2(0) = 1 + 4 - 0 = 5$

EXAMPLE

▶ Evaluate $-3\cos^2 \pi + 2\csc \dfrac{3\pi}{2} + \tan 0$. Note: $\cos^2 \pi = (\cos \pi)^2$.

$-3\cos^2 \pi + 2\csc \dfrac{3\pi}{2} + \tan 0 = -3(-1)^2 + 2(-1) + (0) = -3 - 2 + 0 = -5$

EXERCISE 6-4

Evaluate as indicated.

1. $\sin 90° + \cos 180°$

2. $\csc \dfrac{\pi}{2} - \sin \dfrac{3\pi}{2}$

3. $5\sec^2 \pi + 2\tan \pi - \sec \pi$

4. $4\sin^2 90° - \tan 180° + 3\cos^2 90°$

5. $-8\csc^2 \dfrac{3\pi}{2} + \cos \dfrac{\pi}{2} - \sec 0$

6. $\tan \pi - 2\cos \pi + 3\csc \dfrac{3\pi}{2} + \sin \dfrac{\pi}{2}$

7. $5\sec 180° - 4\cos 270° + \sin 360° + 3\cot 90°$

8. $4\cos 2\pi + 3\sin \pi - 3\cos \pi + \sin \dfrac{\pi}{2}$

9. $\sin 0 + 2\cos 0 + 3\sin \dfrac{\pi}{2} + 4\cos \dfrac{\pi}{2} + 5\sec 0$
$+ 6\csc \dfrac{\pi}{2}$

10. $4\cos 270° - 5\sec 180° - 6\csc 270° + \sin 180°$
$+ 2\cos 180° + 3\sin 270°$

Trigonometric Functions of Coterminal Angles

The values of the trigonometric functions remain the same if an angle is replaced by one that is coterminal with the angle (refer back to Chapter 1 for a thorough discussion of coterminal angles). If an angle is greater than 360° (or 2π) or is negative, you can find an equivalent nonnegative coterminal angle that is less than 360° (or 2π) by adding or subtracting a positive integer multiple of 360° (or 2π).

EXAMPLE

Find the exact value of $\cos 1140°$.

Because $1140° - (3 \times 360°) = 1140° - 1080° = 60°$,

$$\cos 1140° = \cos 60° = \frac{1}{2}.$$

EXAMPLE

Find the exact value of $\sin\left(-\frac{7\pi}{2}\right)$.

Because $-\frac{7\pi}{2} + (2 \times 2\pi) = -\frac{7\pi}{2} + 4\pi = \frac{\pi}{2}$, $\sin\left(-\frac{7\pi}{2}\right) = \sin\frac{\pi}{2} = 1.$

EXERCISE 6-5

For questions 1 to 10, express the trigonometric function as a function of a nonnegative angle that is less than 360° or 2π.

1. $\tan 1000°$

2. $\sin(-150°)$

3. $\cos\dfrac{23\pi}{6}$

4. $\sec\left(-\dfrac{5\pi}{4}\right)$

5. $\cot\left(-\dfrac{\pi}{3}\right)$

6. $\csc\dfrac{17\pi}{3}$

7. $\sin(-225°)$

8. $\cos 630°$

9. $\tan\left(-\dfrac{10\pi}{3}\right)$

10. $\sec 1485°$

For questions 11 to 20, replace the given angle by a nonnegative coterminal angle that is less than 360° or 2π, and then find the exact value of the trigonometric expression.

11. $\cos(-1410°)$

12. $\sec 1485°$

13. $\sin\left(-\dfrac{5\pi}{3}\right)$

14. $\csc 450°$

15. $\tan\dfrac{13\pi}{4}$

16. $5\sqrt{3}\tan\left(-\dfrac{11\pi}{6}\right)$

17. $6\sin 750° - 2\cos 780°$

18. $3\sqrt{2}\sin(-675°) + 2\sqrt{3}\cos(-690°)$

19. $-\tan\dfrac{9\pi}{4}\sin\dfrac{13\pi}{6}$

20. $\cos\left(-\dfrac{17\pi}{3}\right)\sin\left(-\dfrac{17\pi}{3}\right)$

Trigonometric Functions of Negative Angles

Given an angle θ, trigonometric functions for which the trigonometric value of $-\theta$ equals the trigonometric value of θ are **even functions**, and those for which the trigonometric value of $-\theta$ equals the negative of the trigonometric value of θ are **odd functions**. As shown in the table below, cosine and secant are even trigonometric functions and sine, cosecant, tangent, and cotangent are odd trigonometric functions.

Even Functions	Odd Functions
$\cos(-\theta) = \cos\theta$	$\sin(-\theta) = -\sin\theta$
$\sec(-\theta) = \sec\theta$	$\csc(-\theta) = -\csc\theta$
	$\tan(-\theta) = -\tan\theta$
	$\cot(-\theta) = -\cot\theta$

Look at the following:

$$\sin(-75°) = -\sin 75°$$

$$\tan\left(-\frac{\pi}{4}\right) = -\tan\frac{\pi}{4}$$

$$\cos(-150°) = \cos 150°$$

If $\sin(-\theta) = \dfrac{4}{5}$, then $\sin\theta = -\dfrac{4}{5}$.

If $\sec(-\theta) = 2$, then $\cos\theta = \dfrac{1}{2}$.

Because 330° and −30° are coterminal, $\tan(330°) = \tan(-30°) = -\tan 30° = -\dfrac{\sqrt{3}}{3}$.

EXERCISE 6-6

For questions 1 to 5, express the trigonometric function as a function of an acute angle.

1. $\sin 350°$

2. $\tan\dfrac{23\pi}{12}$

3. $\cot 295°$

4. $\sec 312°$

5. $\cos\dfrac{16\pi}{9}$

For questions 6 to 10, solve as indicated.

6. Given $\tan(-\theta) = \dfrac{1}{3}$. Find $\cot(\theta)$.

7. Given $\sin(-\theta) = \dfrac{11}{60}$. Find $\csc(\theta)$.

8. Given $\sec(-\theta) = -\dfrac{25}{7}$. Find $\cos(\theta)$.

9. Given $\cot(-\theta) = -\dfrac{3}{4}$. Find $\tan(\theta)$.

10. Given $\cos(-\theta) = \dfrac{48}{73}$. Find $\sec(\theta)$.

For questions 11 to 20, find the exact value of the given expression.

11. $\cos 300°$

12. $\sin \dfrac{11\pi}{6}$

13. $\tan 315°$

14. $\sin (-750°)$

15. $\cot \dfrac{7\pi}{4}$

16. $3\sqrt{2}\sin 1035°$

17. $4\sin(-30°) + 2\cos(-30°)$

18. $5\sqrt{2}\tan(-45°) + 4\cos(-60°)$

19. $4\sin(-60°)\cos(-30°)$

20. $2\sqrt{3}\tan\left(-\dfrac{\pi}{3}\right) + 10\tan\left(-\dfrac{\pi}{4}\right)$

Using Reference Angles to Find the Values of Trigonometric Functions

A trigonometric function value of an angle has the same absolute value as the trigonometric function value of its reference angle (refer back to Chapter 1 for a thorough discussion of reference angles). The sign (either positive or negative) of the function value depends on the quadrant of the original angle as shown in the following table.

The relationship of a positive angle θ that is less than 360° and its reference angle θ_r in each quadrant is given in the following table.

θ's Quadrant	θ_r	Signs of Functions
I	$\theta_r = \theta$	All functions are positive.
II	$\theta_r = 180° - \theta$	Only $\sin \theta$ and $\csc \theta$ are positive.
III	$\theta_r = \theta - 180°$	Only $\tan \theta$ and $\cot \theta$ are positive.
IV	$\theta_r = 360° - \theta$	Only $\cos \theta$ and $\sec \theta$ are positive.

If θ is a quadrantal angle, then the values of the trigonometric functions for θ are the same as given earlier in this chapter, so a reference angle is not needed.

EXAMPLE

▶ Express sin 235° as the sine of an acute angle.

$\sin 235° = -\sin (235° - 180°) = -\sin 55°$

EXAMPLE

▶ Express $\tan\dfrac{3\pi}{5}$ as the tangent of an acute angle.

$$\tan\dfrac{3\pi}{5} = -\tan\left(\pi - \dfrac{3\pi}{5}\right) = -\tan\dfrac{2\pi}{5}$$

When the reference angle for a trigonometric function is a special acute angle, you can determine the exact value of the trigonometric function (refer back to Chapter 3 for a thorough discussion of special acute angles).

EXAMPLE

▶ Find the exact value of $\sin\dfrac{5\pi}{6}$.

$$\sin\dfrac{5\pi}{6} = \sin\left(\pi - \dfrac{5\pi}{6}\right) = \sin\dfrac{\pi}{6} = \dfrac{1}{2}$$

Values of the trigonometric functions for special angles and the quadrantal angles from 0° to 360° are given in the following table.

Trigonometric Function Values of Special Angles and Quadrantal Angles						
θ (Deg or Rad)	$\sin\theta$	$\cos\theta$	$\tan\theta$	$\sec\theta$	$\csc\theta$	$\cot\theta$
0° or 0	0	1	0	1	undefined	undefined
30° or $\dfrac{\pi}{6}$	$\dfrac{1}{2}$	$\dfrac{\sqrt{3}}{2}$	$\dfrac{\sqrt{3}}{3}$	$\dfrac{2\sqrt{3}}{3}$	2	$\sqrt{3}$
45° or $\dfrac{\pi}{4}$	$\dfrac{\sqrt{2}}{2}$	$\dfrac{\sqrt{2}}{2}$	1	$\sqrt{2}$	$\sqrt{2}$	1
60° or $\dfrac{\pi}{3}$	$\dfrac{\sqrt{3}}{2}$	$\dfrac{1}{2}$	$\sqrt{3}$	2	$\dfrac{2\sqrt{3}}{3}$	$\dfrac{\sqrt{3}}{3}$
90° or $\dfrac{\pi}{2}$	1	0	undefined	undefined	1	0
120° or $\dfrac{2\pi}{3}$	$\dfrac{\sqrt{3}}{2}$	$-\dfrac{1}{2}$	$-\sqrt{3}$	-2	$\dfrac{2\sqrt{3}}{3}$	$-\dfrac{\sqrt{3}}{3}$
135° or $\dfrac{3\pi}{4}$	$\dfrac{\sqrt{2}}{2}$	$-\dfrac{\sqrt{2}}{2}$	-1	$-\sqrt{2}$	$\sqrt{2}$	-1
150° or $\dfrac{5\pi}{6}$	$\dfrac{1}{2}$	$-\dfrac{\sqrt{3}}{2}$	$-\dfrac{\sqrt{3}}{3}$	$-\dfrac{2\sqrt{3}}{3}$	2	$-\sqrt{3}$
180° or π	0	-1	0	-1	undefined	undefined

Continued

$210°$ or $\dfrac{7\pi}{6}$	$-\dfrac{1}{2}$	$-\dfrac{\sqrt{3}}{2}$	$\dfrac{\sqrt{3}}{3}$	$-\dfrac{2\sqrt{3}}{3}$	-2	$\sqrt{3}$
$225°$ or $\dfrac{5\pi}{4}$	$-\dfrac{\sqrt{2}}{2}$	$-\dfrac{\sqrt{2}}{2}$	1	$-\sqrt{2}$	$-\sqrt{2}$	1
$240°$ or $\dfrac{4\pi}{3}$	$-\dfrac{\sqrt{3}}{2}$	$-\dfrac{1}{2}$	$\sqrt{3}$	-2	$-\dfrac{2\sqrt{3}}{3}$	$\dfrac{\sqrt{3}}{3}$
$270°$ or $\dfrac{3\pi}{2}$	-1	0	undefined	undefined	-1	0
$300°$ or $\dfrac{5\pi}{3}$	$-\dfrac{\sqrt{3}}{2}$	$\dfrac{1}{2}$	$-\sqrt{3}$	2	$-\dfrac{2\sqrt{3}}{3}$	$-\dfrac{\sqrt{3}}{3}$
$315°$ or $\dfrac{7\pi}{4}$	$-\dfrac{\sqrt{2}}{2}$	$\dfrac{\sqrt{2}}{2}$	-1	$\sqrt{2}$	$-\sqrt{2}$	-1
$330°$ or $\dfrac{11\pi}{6}$	$-\dfrac{1}{2}$	$\dfrac{\sqrt{3}}{2}$	$-\dfrac{\sqrt{3}}{3}$	$\dfrac{2\sqrt{3}}{3}$	-2	$-\sqrt{3}$
$360°$ or 2π	0	1	0	1	undefined	undefined

You frequently will encounter these angles in your study of trigonometry. Given your understanding of reference angles, if you know the function values of $0°$, $30°$, $45°$, $60°$, and $90°$, and if you know the signs of the trigonometric functions in each of the four quadrants, then you can reproduce the table above without the use of any other aid. For most other angles, a calculator (or similar resource) is needed to determine the values of their trigonometric functions.

The algebraic sign of the three basic trigonometric functions in each quadrant is easily remembered by using the mnemonic: "**A**ll **s**tudents **t**ake **c**alculus," which reminds you that "**A**ll three are positive in quadrant I; the **s**ine is positive in quadrant II; the **t**angent is positive in quadrant III; and the **c**osine is positive in quadrant IV.

EXERCISE 6-7

For questions 1 to 15, express the trigonometric function as a function of an acute angle.

1. $\cos 125°$

2. $\cot \dfrac{9\pi}{5}$

3. $\sin 912°$

4. $\csc (-100°)$

5. $\tan (-227°)$

6. $\sin 325°$

7. $\cot 680°$

8. $\cos \dfrac{10\pi}{9}$

9. $\tan (-290°)$

10. $\csc (-680°)$

11. $\cos 240°$

12. $\tan 120°$

13. $\sin \left(-\dfrac{7\pi}{4}\right)$

14. $\cos \left(-\dfrac{\pi}{8}\right)$

15. $\sin (-315°)$

For questions 16 to 25, find the exact value of the given expression.

16. $\sin\dfrac{\pi}{2}$

17. $\tan\dfrac{5\pi}{6}$

18. $\sec\pi$

19. $\cot 120°$

20. $\sin(-315°)$

21. $\sec 135°$

22. $\cos\dfrac{3\pi}{4} + \sin\left(-\dfrac{\pi}{4}\right)$

23. $\tan 60° - \cot(-30°)$

24. $\sin\dfrac{3\pi}{4}\cos\dfrac{5\pi}{3}$

25. $\tan 150° \tan 225°$

Trigonometric Identities

Definition and Guidelines

A **trigonometric identity** is an equation that is true for all values of the variable(s) in the domain(s) of the associated trigonometric function(s). (See Appendix B for a list of identities.) An underlying assumption is that no expression in any denominator is equal to 0. Otherwise, the expression is undefined.

In this workbook, the terms *identity* and *formula* have the same meaning.

EXAMPLE

$\cos\theta\tan\theta = \sin\theta$ is a trigonometric identity because it is true for all values of θ in the domains of $\cos\theta, \tan\theta,$ and $\sin\theta$.

You verify an identity by transforming one side of the identity until it is identical to the other side. Here are some guidelines to keep in mind:

▶ Work with only one side of the identity. It doesn't matter which side you start with, but as a general rule, start with the side that appears to be more complicated.

▶ Do not treat the identity as an equation. You cannot assume it is true, so equality properties cannot be applied.

▶ Converting all functions to sines and cosines is often helpful.

▶ Use additional trigonometric identities (such as the ones you will encounter in subsequent lessons in this chapter).

▶ Use algebraic manipulation (factoring, combining fractional expressions, multiplying/squaring expressions, and so forth).

Note to the reader: The first two bullets in the guidelines above are nonnegotiable. However, within those constraints, there is often room for flexibility when you are verifying identities. When looking at the examples and working through the identities that follow, keep in mind there can be multiple ways to transform one side of an identity so that it is identical to the other side. You might think of a way to accomplish that goal other than what is shown.

EXAMPLE

Proving or verifying an identity *cannot* be accomplished by repeatedly substituting in domain values to obtain a true equation. Proving identities *can* be accomplished by the use of good logic, referencing fundamental identities, and using algebraic skills and substitution principles.

▶ Verify the identity: $\cos\theta\tan\theta = \sin\theta$

▶ The approach is to logically transform one side of the identity into the other side. Starting with the left side, proceed as follows:

$$\cos\theta\tan\theta = \cos\theta\frac{\sin\theta}{\cos\theta} \quad \left(\begin{array}{c}\text{by the ratio identity for tangent given}\\ \text{later in this chapter}\end{array}\right)$$

$$= \cancel{\cos\theta}\,\frac{\sin\theta}{\cancel{\cos\theta}}$$

$$= \sin\theta$$

▶ Therefore, $\cos\theta\tan\theta = \sin\theta$ is an identity.

EXAMPLE

▶ Verify that $\sin(\theta + \beta) = \sin\theta + \sin\beta$ is *not* an identity.

▶ It takes only one counterexample to establish that an equation is not an identity. For that purpose, you are permitted to substitute in domain values to show the statement is false.

Let $\theta = \beta = \dfrac{\pi}{3}$. Then, on the left side, you have

$$\sin(\theta + \beta) = \sin\left(\frac{2\pi}{3}\right) = \frac{\sqrt{3}}{2}.\text{ But, on the right side,}$$

$$\sin\theta + \sin\beta = \sin\frac{\pi}{3} + \sin\frac{\pi}{3} = \frac{\sqrt{3}}{2} + \frac{\sqrt{3}}{2} = \sqrt{3}.\text{ Because }\frac{\sqrt{3}}{2} \neq \sqrt{3},$$

the equation is not an identity.

EXERCISE 7-1

Fill in the blank to make a true statement.

1. A trigonometric identity is an equation that is true for all values of the _____ in the domains of the associated trigonometric functions.

2. An underlying assumption of a trigonometric identity is that no expression in any denominator is equal to _____.

3. You verify an identity by transforming one side of the identity until it is _____ to the other side.

4. It takes only one _____ to establish that an equation is not an identity.

5. Verifying an identity _____ (can, cannot) be accomplished by repeatedly substituting in domain values to obtain a true equation.

The Reciprocal and Ratio Identities

The reciprocal identities and ratio identities are defined as follows:

▶ The **reciprocal identities**:

$$\sec\theta = \frac{1}{\cos\theta} \qquad \csc\theta = \frac{1}{\sin\theta} \qquad \cot\theta = \frac{1}{\tan\theta}.$$

▶ The **ratio identities**:

$$\tan\theta = \frac{\sin\theta}{\cos\theta} \qquad \cot\theta = \frac{\cos\theta}{\sin\theta}$$

Every trigonometric function can be written in terms of the sine and cosine functions.

EXAMPLE

▶ Verify the identity: $\sec\theta(\cos\theta + 1) = 1 + \sec\theta$.

$$\sec\theta(\cos\theta + 1) = \frac{1}{\cos\theta}(\cos\theta + 1)$$
$$= \frac{\cos\theta}{\cos\theta} + \frac{1}{\cos\theta}$$
$$= 1 + \sec\theta$$

▶ Therefore, $\sec\theta(\cos\theta + 1) = 1 + \sec\theta$ is an identity.

EXAMPLE

Writing an expression in terms of a single function can be useful for numerical work. To get a numerical value, you would have to evaluate only one function rather than two or more individually.

Use identities to express $\dfrac{\sin^4 \theta}{\cos^4 \theta}$ in terms of a single function.

$$\frac{\sin^4 \theta}{\cos^4 \theta} = \left(\frac{\sin\theta}{\cos\theta}\right)^4 = \tan^4 \theta$$

EXERCISE 7-2

For questions 1 to 6, use identities to write the expression in terms of a single function.

1. $\dfrac{\sin^3 \theta}{\cos^3 \theta}$

2. $\dfrac{\sec\theta}{\tan\theta}$

3. $\sec \theta \cot \theta$

4. $\csc \theta \tan \theta$

5. $\dfrac{\csc\theta}{\cot\theta}$

6. $\dfrac{\cot^2 \theta}{\csc^2 \theta}$

For questions 7 to 12, verify that the statement is an identity.

7. $\dfrac{\sin^2 2\theta \cot 2\theta}{\cos 2\theta} = \sin 2\theta$

8. $(\sin\theta)(\cot^2 \theta)(\sec^2 \theta) = \csc\theta$

9. $\dfrac{\sec\theta}{\cot\theta} = \dfrac{\sin\theta}{\cos^2 \theta}$

10. $\dfrac{\cot\theta}{\csc\theta} = \cos\theta$

11. $\dfrac{\cot^2 \theta}{\sec^2 \theta} = \dfrac{\cos^4 \theta}{\sin^2 \theta}$

12. Express $\sec\theta + \tan\theta$ as a single fraction.

For questions 13 to 15, indicate whether the statement is True or False.

13. $\dfrac{\sin 2\theta}{\cos 4\theta} = \dfrac{\sin\theta}{\cos 2\theta}$

14. $\dfrac{\sin^2 \theta}{\sin^4 \theta} = \dfrac{1}{\sin^2 \theta}$

15. $\sec\theta + \csc\theta = \dfrac{1}{\cos\theta + \sin\theta}$

The Pythagorean Identities

The following Pythagorean identities can be obtained from the trigonometric ratios discussed in the first section of Chapter 6.

$$\sin^2 \theta + \cos^2 \theta = 1 \qquad \tan^2 \theta + 1 = \sec^2 \theta \qquad \cot^2 \theta + 1 = \csc^2 \theta$$

EXAMPLE

Consider the following figure.

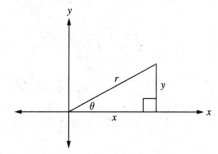

From the definitions of the trigonometric ratios, $\sin\theta = \dfrac{y}{r}$ and

$\cos\theta = \dfrac{x}{r}$. It follows that $\sin^2\theta + \cos^2\theta = \left(\dfrac{y}{r}\right)^2 + \left(\dfrac{x}{r}\right)^2 = \dfrac{y^2 + x^2}{r^2}$.

By the Pythagorean theorem, $y^2 + x^2 = r^2$. Thus, $\sin^2\theta + \cos^2\theta = 1$.

EXAMPLE

Verify that $\tan^2\theta + 1 = \sec^2\theta$ is an identity.

$$\tan^2\theta + 1 = \left(\frac{\sin\theta}{\cos\theta}\right)^2 + 1$$

$$= \frac{\sin^2\theta}{\cos^2\theta} + 1$$

$$= \frac{\sin^2\theta + \cos^2\theta}{\cos^2\theta}$$

$$= \frac{1}{\cos^2\theta}$$

$$= \sec^2\theta$$

Hence, $\tan^2\theta + 1 = \sec^2\theta$ is an identity.

EXERCISE 7-3

For questions 1 and 2, answer as indicated.

1. Express $\dfrac{\sin\theta}{\cos\theta} + \dfrac{\cos\theta}{\sin\theta}$ as a single fraction.

2. Express $\dfrac{\csc\theta}{\sec\theta} + \dfrac{\sec\theta}{\csc\theta}$ as a single fraction in terms of $\sin\theta$ and $\cos\theta$.

For questions 3 to 15, verify the identity.

3. $\cot^2\theta + 1 = \csc^2\theta$

4. $\csc\theta + \sin\theta = \dfrac{\sin^2\theta + 1}{\sin\theta}$

5. $1 + \tan^2\theta = \dfrac{1}{\cos^2\theta}$

6. $\cos^4\theta - \sin^4\theta = \sin^2\theta - \cos^2\theta$

7. $\cos^2\theta(1 + \tan^2\theta) = 1$

8. $\dfrac{\sin^2\theta}{1 - \cos\theta} = \dfrac{1 + \sec\theta}{\sec\theta}$

9. $\dfrac{1}{1 - \sin\theta} + \dfrac{1}{1 + \sin\theta} = \dfrac{2}{\cos^2\theta}$

10. $(\csc^2\theta - 1)(\sec^2\theta - 1) = 1$

11. $\dfrac{1 + \tan^2\theta}{\csc^2\theta} = \tan^2\theta$

12. $\dfrac{\sin^2\theta}{\tan^2\theta - \sin^2\theta} = \cot^2\theta$

13. $\cos^2\theta(1 + \tan^2\theta) = 1$

14. $\tan^2\theta + \sec^2\theta = \dfrac{2}{\cos^2\theta} - 1$

15. $\dfrac{1 + \tan^2\theta}{\tan^2\theta} = \csc^2\theta$

For questions 16 to 20, indicate whether the statement is True or False.

16. $\sec^2\theta + \csc^2\theta = 1$

17. $\sec\theta + \cos\theta = \dfrac{1 + \cos^2\theta}{\cos\theta}$

18. $\tan\theta - \cot\theta\tan^2\theta = 0$

19. $\sin 3\theta - \sin 2\theta = \sin\theta$

20. $\dfrac{\cos^2\theta + \tan^2\theta - 1}{\tan^2\theta} = \sin^2\theta$

Sum and Difference Formulas for the Sine Function

The following are the sum and difference formulas for the sine function.

$$\sin(\theta + \varphi) = \sin\theta\cos\varphi + \cos\theta\sin\varphi$$
$$\sin(\theta - \varphi) = \sin\theta\cos\varphi - \cos\theta\sin\varphi$$

EXAMPLE

Find the exact value of $\sin 15°$.

$$\sin 15° = \sin(45° - 30°) = \sin 45°\cos 30° - \cos 45°\sin 30°$$

$$= \frac{\sqrt{2}}{2}\frac{\sqrt{3}}{2} - \frac{\sqrt{2}}{2}\left(\frac{1}{2}\right) = \frac{\sqrt{6} - \sqrt{2}}{4}$$

EXAMPLE

Find the exact value of $\sin\dfrac{\pi}{12}$.

$$\sin\frac{\pi}{12} = \sin\left(\frac{\pi}{3} - \frac{\pi}{4}\right) = \sin\frac{\pi}{3}\cos\frac{\pi}{4} - \cos\frac{\pi}{3}\sin\frac{\pi}{4}$$

$$= \frac{\sqrt{3}}{2}\left(\frac{\sqrt{2}}{2}\right) - \frac{1}{2}\left(\frac{\sqrt{2}}{2}\right) = \frac{\sqrt{6} - \sqrt{2}}{4}$$

EXAMPLE

Write $\sin 5\theta\cos 3\theta + \cos 5\theta\sin 3\theta$ as a single function of $k\theta$ for some integer k.

$$\sin 5\theta\cos 3\theta + \cos 5\theta\sin 3\theta = \sin(5\theta + 3\theta) = \sin 8\theta$$

EXAMPLE

Verify the identity: $\sin(180° - \theta) = \sin\theta$

$$\sin(180° - \theta) = \sin 180°\cos\theta - \cos 180°\sin\theta = 0 \cdot \cos\theta - (-1)\sin\theta = \sin\theta$$

EXERCISE 7-4

For questions 1 to 8, find the exact value of the expression.

1. $\sin 105°$

2. $\sin 195°$

3. $\sin 75°$

4. $\sin 165°$

5. $\sin\dfrac{5\pi}{12}$

6. $\sin\dfrac{7\pi}{12}$

7. $\sin 180°$

8. $\sin 120°$

For questions 9 and 10, evaluate the given expression without using a calculator.

9. $\sin 20°\cos 160° + \cos 20°\sin 160°$

10. $\sin 110°\cos 80° - \cos 110°\sin 80°$

For questions 11 to 20, write the expression as a single function of $k\theta$ for some integer k.

11. $\sin 5\theta\cos\theta + \cos 5\theta\sin\theta$

12. $\sin\theta\cos 2\theta + \cos\theta\sin 2\theta$

13. $\sin 10\theta\cos 6\theta - \cos 10\theta\sin 6\theta$

14. $\sin\dfrac{3}{4}\theta\cos\dfrac{5}{4}\theta + \cos\dfrac{3}{4}\theta\sin\dfrac{5}{4}\theta$

15. $\sin 5\theta\cos 4\theta - \cos 5\theta\sin 4\theta$

16. $\sin\pi\theta\cos(\pi - 4)\theta - \cos\pi\theta\sin(\pi - 4)\theta$

17. $\sin 8\theta\cos 6\theta + \cos 8\theta\sin 6\theta$

18. $\sin\dfrac{2}{3}\theta\cos\dfrac{1}{3}\theta + \cos\dfrac{2}{3}\theta\sin\dfrac{1}{3}\theta$

19. $\sin(\theta + 70°)\cos(\theta - 70°) + \cos(\theta + 70)\sin(\theta - 70°)$

20. $\sin(\theta + 3\pi^2)\cos(\theta - 3\pi^2) + \cos(\theta + 3\pi^2)\sin(\theta - 3\pi^2)$

For questions 21 to 25, verify the identity.

21. $\sin(180° + \theta) = -\sin\theta$

22. $\sin(360° - \theta) = -\sin\theta$

23. $\sin(90° - \theta) = \cos\theta$

24. $\sin\left(\dfrac{3\pi}{2} + \theta\right) = -\cos\theta$

25. $\sin\left(\theta - \dfrac{\pi}{6}\right) = \dfrac{\sqrt{3}\sin\theta - \cos\theta}{2}$

Sum and Difference Formulas for the Cosine Function

The following are the sum and difference formulas for the cosine function.

$$\cos(\theta + \phi) = \cos\theta\cos\phi - \sin\theta\sin\phi$$
$$\cos(\theta - \phi) = \cos\theta\cos\phi + \sin\theta\sin\phi$$

EXAMPLE

▶ Find the exact value of cos 75°.

$$\cos 75° = \cos(45° + 30°) = \cos 45° \cos 30° - \sin 45° \sin 30°$$

$$= \frac{\sqrt{2}}{2}\left(\frac{\sqrt{3}}{2}\right) - \frac{\sqrt{2}}{2}\left(\frac{1}{2}\right) = \frac{\sqrt{6} - \sqrt{2}}{4}$$

EXAMPLE

▶ Find the exact value of $\cos\dfrac{\pi}{12}$.

$$\cos\frac{\pi}{12} = \cos\left(\frac{\pi}{4} - \frac{\pi}{6}\right) = \cos\frac{\pi}{4}\cos\frac{\pi}{6} + \sin\frac{\pi}{4}\sin\frac{\pi}{6}$$

$$= \frac{\sqrt{2}}{2}\left(\frac{\sqrt{3}}{2}\right) + \frac{\sqrt{2}}{2}\left(\frac{1}{2}\right) = \frac{\sqrt{6} + \sqrt{2}}{4}$$

EXAMPLE

Write $\cos 2\theta \cos 3\theta - \sin 2\theta \sin 3\theta$ as a single function of $k\theta$ for some integer k.

$$\cos 2\theta \cos 3\theta - \sin 2\theta \sin 3\theta = \cos(2\theta + 3\theta) = \cos 5\theta$$

EXERCISE 7-5

For questions 1 to 8, find the exact value of the expression.

1. $\cos 105°$

2. $\cos 195°$

3. $\cos \dfrac{5\pi}{12}$

4. $\cos 165°$

5. $\cos\left(\dfrac{\pi}{3} - \dfrac{\pi}{4}\right)$

6. $\cos(180°)$

7. $\cos(120°)$

8. $\cos\left(\dfrac{4\pi}{3}\right)$

For questions 9 and 10, evaluate the given expression without using a calculator.

9. $\cos 130° \cos 70° + \sin 130° \sin 70°$

10. $\cos 35° \cos 55° - \sin 35° \sin 55°$

For questions 11 to 20, write the expression as a single function of $k\theta$ for some integer k.

11. $\cos 5\theta \cos \theta + \sin 5\theta \sin \theta$

12. $\cos \theta \cos 2\theta + \sin \theta \sin 2\theta$

13. $\cos \theta \cos \theta - \sin \theta \sin \theta$

14. $\cos \dfrac{3}{4}\theta \cos \dfrac{5}{4}\theta - \sin \dfrac{3}{4}\theta \sin \dfrac{5}{4}\theta$

15. $\cos 5\theta \cos 4\theta + \sin 5\theta \sin 4\theta$

16. $\cos \pi\theta \cos(\pi - 4)\theta + \sin \pi\theta \sin(\pi - 4)\theta$

17. $\cos 8\theta \cos 6\theta + \sin 8\theta \sin 6\theta$

18. $\cos \dfrac{2}{3}\theta \cos \dfrac{1}{3}\theta - \sin \dfrac{2}{3}\theta \sin \dfrac{1}{3}\theta$

19. $\cos(\theta + 70°)\cos(\theta - 70°) - \sin(\theta + 70°)\sin(\theta - 70°)$

20. $\cos(\theta + 3\pi^2)\cos(\theta - 3\pi^2) - \sin(\theta + 3\pi^2)\sin(\theta - 3\pi^2)$

For questions 21 to 25, verify the identity.

21. $\cos(180° - \theta) = -\cos\theta$

22. $\cos(180° + \theta) = -\cos\theta$

23. $\cos(360° - \theta) = \cos\theta$

24. $\cos\left(\dfrac{5\pi}{2} + \theta\right) = -\sin\theta$

25. $\cos\left(\theta - \dfrac{\pi}{3}\right) = \dfrac{\cos\theta + \sqrt{3}\sin\theta}{2}$

Sum and Difference Formulas for the Tangent Function

The sum and difference formulas for the tangent function are derived from those for the sine and cosine functions. The derivation follows as an example of using identities to gain further information.

$$
\begin{aligned}
\tan(\theta + \varphi) &= \frac{\sin(\theta + \varphi)}{\cos(\theta + \varphi)} \\
&= \frac{\sin\theta\cos\varphi + \cos\theta\sin\varphi}{\cos\theta\cos\varphi - \sin\theta\sin\varphi} \\
&= \frac{\cos\theta\cos\varphi\left(\dfrac{\sin\theta}{\cos\theta} + \dfrac{\sin\varphi}{\cos\varphi}\right)}{\cos\theta\cos\varphi\left(1 - \dfrac{\sin\theta\sin\varphi}{\cos\theta\cos\varphi}\right)} \\
&= \frac{\tan\theta + \tan\varphi}{1 - \tan\theta\tan\varphi}
\end{aligned}
$$

Thus, you have the following formula:

$$
\tan(\theta + \varphi) = \frac{\tan\theta + \tan\varphi}{1 - \tan\theta\tan\varphi}
$$

And by a similar computation:

$$
\tan(\theta - \varphi) = \frac{\tan\theta - \tan\varphi}{1 + \tan\theta\tan\varphi}
$$

EXAMPLE

Find the exact value of tan 15°.

$$
\tan 15° = \tan(45° - 30°) = \frac{\tan 45° - \tan 30°}{1 + \tan 45° \tan 30°} = \frac{1 - \dfrac{1}{\sqrt{3}}}{1 + \dfrac{1}{\sqrt{3}}} = \frac{\sqrt{3} - 1}{\sqrt{3} + 1}
$$

EXAMPLE

Verify that $\dfrac{\tan 65° - \tan 20°}{1 + \tan 65° \tan 20°} = 1$

$$\dfrac{\tan 65° - \tan 20°}{1 + \tan 65° \tan 20°} = \tan(65° - 20°) = \tan 45° = 1$$

EXAMPLE

Write the expression $\dfrac{\tan 2\theta + \tan\theta}{1 - \tan 2\theta \tan\theta}$ as a single function of $k\theta$ where k is an integer.

$$\dfrac{\tan 2\theta + \tan\theta}{1 - \tan 2\theta \tan\theta} = \tan\big(2\theta + \theta\big) = \tan 3\theta$$

EXERCISE 7-6

For questions 1 to 5, find the exact value of the given angle.

1. $\tan 195°$

2. $\tan 75°$

3. $\tan 105°$

4. $\tan\dfrac{\pi}{12}$

5. $\tan 165°$

For questions 6 to 10, verify the statement is correct.

6. $\dfrac{\tan 130° + \tan 50°}{1 - \tan 130° \tan 50°} = 0$

7. $\dfrac{\tan 110° - \tan 50°}{1 + \tan 110° \tan 50°} = \sqrt{3}$

8. $\dfrac{\tan 115° - \tan 70°}{1 + \tan 115° \tan 70°} = 1$

9. $\dfrac{\tan 100° + \tan 50°}{1 - \tan 100° \tan 50°} = -\dfrac{1}{\sqrt{3}}$

10. $\dfrac{\tan 95° + \tan 40°}{1 - \tan 95° \tan 40°} = -1$

For questions 11 to 15, write the expression as a single function of $k\theta$ where k is an integer.

11. $\dfrac{\tan 4\theta + \tan 5\theta}{1 - \tan 4\theta \tan 5\theta}$

12. $\dfrac{\tan 5\theta - \tan\theta}{1 + \tan 5\theta \tan\theta}$

13. $\dfrac{\tan 8\theta + \tan 6\theta}{1 - \tan 8\theta \tan 6\theta}$

14. $\dfrac{\tan\dfrac{4}{3}\theta - \tan\dfrac{1}{3}\theta}{1 + \tan\dfrac{4}{3}\theta \tan\dfrac{1}{3}\theta}$

15. $\dfrac{\tan\dfrac{5}{4}\theta + \tan\dfrac{3}{4}\theta}{1 - \tan\dfrac{5}{4}\theta \tan\dfrac{3}{4}\theta}$

For questions 16 and 17, write the expression as a function of an acute angle.

16. $\tan 120°$

17. $\tan 315°$

For questions 18 to 20, verify the equation is correct.

18. $\tan(180° - \theta) = -\tan\theta$

20. $\tan\left(\theta + \dfrac{\pi}{4}\right) = \dfrac{1 + \tan\theta}{1 - \tan\theta}$

19. $\tan 2\theta = \dfrac{2\tan\theta}{1 - \tan^2\theta}$

Reduction Formulas

A reduction formula is an identity used to reduce the complexity of a trigonometric function.

Reduction formulas for the sine, cosine, and tangent functions are the following:

$$\sin(180° - \theta) = \sin\theta \qquad \cos(180° - \theta) = -\cos\theta \qquad \tan(180° - \theta) = -\tan\theta$$
$$\sin(180° + \theta) = -\sin\theta \qquad \cos(180° + \theta) = -\cos\theta \qquad \tan(180° + \theta) = \tan\theta$$
$$\sin(360° - \theta) = -\sin\theta \qquad \cos(360° - \theta) = \cos\theta \qquad \tan(360° - \theta) = -\tan\theta$$

EXAMPLE

Solve for θ: $\sin(180° - 4\theta) = \sin 80°$ for $0 \le \theta \le 90°$.

Using the reduction formula, $\sin(180° - \theta) = \sin\theta$ yields $\sin(180° - 4\theta) = \sin 4\theta = \sin 80°$. It follows that $4\theta = 80°$ or $4\theta = 100°$. Thus, $\theta = 20°$ or $\theta = 25°$.

EXAMPLE

Solve for θ: $\cos(3\theta + 20°) = -\cos(60° - \theta)$ for $0 \le \theta \le 90°$.

Using the reduction formula, $\cos(180° - \theta) = -\cos\theta$, yields $\cos(3\theta + 20°) = -\cos(180° - (60° - \theta))$. It follows that $3\theta + 20° = 180° - (60° - \theta)$. Therefore, $2\theta = 100°$ or $\theta = 50°$.

EXERCISE 7-7

Solve for θ, where $0 \le \theta \le 90°$.

1. $\sin(180° - 2\theta) = \cos 30°$

4. $\cos(2\theta + 90°) = -\cos 20°$

2. $\tan(180° - 3\theta) = -\tan 120°$

5. $\tan(\theta + 40°) = \tan(2\theta + 5°)$

3. $\sin(180° + 5\theta) = \sin(45° + 2\theta)$

Double-Angle Identities

Using the sum formulas for sine, cosine, and the tangent functions, you can quickly derive the following identities:

$$\sin 2\theta = 2\sin\theta\cos\theta$$
$$\cos 2\theta = \cos^2\theta - \sin^2\theta = 1 - 2\sin^2\theta = 2\cos^2\theta - 1$$
$$\tan 2\theta = \frac{2\tan\theta}{1 - \tan^2\theta}$$

These identities can be used to convert an expression from double angles to single angles or from multiple expressions to a single expression and vice versa.

EXAMPLE

Express $\dfrac{\tan 3\theta}{1 - \tan^2 3\theta}$ as a single function of $k\theta$ where k is an integer.

$$\frac{\tan 3\theta}{1 - \tan^2 3\theta} = \frac{1}{2}\left(\frac{2\tan 3\theta}{1 - \tan^2 3\theta}\right) = \frac{1}{2}\tan 6\theta$$

EXAMPLE

Verify that $\sin 2\theta = \dfrac{2\tan\theta}{1 + \tan^2\theta}$ is an identity.

$$\sin 2\theta = 2\sin\theta\cos\theta = \cos^2\theta\left(2\frac{\sin\theta}{\cos\theta}\right) = \cos^2\theta(2\tan\theta)$$

$$= \frac{1}{\sec^2\theta}(2\tan\theta) = \frac{2\tan\theta}{1 + \tan^2\theta}$$

EXERCISE 7-8

For questions 1 to 14, write the given expression as a single function of $k\theta$ where k is an integer.

1. $\dfrac{4\tan 2\theta}{1 - \tan^2 2\theta}$

2. $\cos^2 5\theta - \sin^2 5\theta$

3. $4\cos^2 6\theta - 2$

4. $2\sin 2\theta\cos 2\theta$

5. $\dfrac{2\tan\theta}{1 - \tan^2\theta}$

6. $\sin\dfrac{\theta}{2}\cos\dfrac{\theta}{2}$

7. $1 - 2\sin^2 3\theta$

8. $\dfrac{\cos 3\theta \sin 3\theta}{12}$

9. $\dfrac{3\tan 2\theta}{5(1 - \tan^2 2\theta)}$

10. $(\sin\theta - \cos\theta)(\sin\theta + \cos\theta)$

11. $\dfrac{\tan^2 4\theta - 1}{6\tan 4\theta}$

12. $\dfrac{\sin 2\theta \cos 2\theta}{\cos^2 2\theta - \sin^2 2\theta}$

13. $\cos^4 \theta - \sin^4 \theta$

14. $\dfrac{\sin 2\theta (6\tan\theta)}{(2 - 2\tan^2 \theta)(2\cos^2 \theta - 1)}$

For questions 15 to 22, verify the identity.

15. $\cos^2 \theta = \dfrac{1 + \cos 2\theta}{2}$ (Hint: Begin

with $\dfrac{1 + \cos 2\theta}{2}$)

16. $\sin 2\theta \csc\theta = 2\cos\theta$

17. $\sec^2 \theta = \dfrac{2}{1 + \cos 2\theta}$

18. $\cot(4\theta) = \dfrac{1 - \tan^2 (2\theta)}{2\tan(2\theta)}$

19. $\csc 2\theta = \dfrac{1}{2}\cot\theta + \dfrac{1}{2}\tan\theta$

20. $\sin 4\theta = 4\cos^3 \theta \sin\theta - 4\cos\theta \sin^3 \theta$

21. $\dfrac{\tan 3\theta}{1 - \tan^2 3\theta} = \dfrac{1}{2}\tan 6\theta$

22. $\cos 3\theta = 4\cos^3 \theta - 3\cos\theta$ (Hint: $3\theta = 2\theta + \theta$)

For questions 23 to 27, find the exact value of sin 2θ and cos 2θ without using a calculator.

23. $\cos\theta = -\dfrac{12}{13}$, θ in quadrant III

24. $\tan\theta = \dfrac{7}{24}$, θ in quadrant III

25. $\sin\theta = -\dfrac{8}{17}$, θ in quadrant IV

26. $\sec\theta = \sqrt{2}$, θ in quadrant I

27. $\cos\theta = -\dfrac{3}{5}$, θ in quadrant II

For questions 28 to 30, indicate whether the statement is True or False.

28. $\sin 78° = \dfrac{6\sqrt{3}\cos 39° \sin 39°}{3}\left(\dfrac{1}{\sqrt{3}}\right)$

29. $\dfrac{\tan 3\theta\left(1 - \tan^2 \dfrac{3\theta}{2}\right)}{2} = \tan\dfrac{3\theta}{2}$

30. $\sin^2 \theta - \cos^2 \theta = \cos(-2\theta)$

Half-Angle Identities

As shown below, you can rearrange the double-angle cosine formulas, $\cos 2\theta = 2\cos^2\theta - 1$ and $\cos 2\theta = 1 - 2\sin^2\theta$, to develop the half-angle identities.

First, transform the two equations as follows:

$$\cos^2\theta = \frac{1 + \cos 2\theta}{2}$$

$$\sin^2\theta = \frac{1 - \cos 2\theta}{2}$$

Next, substitute $\frac{\theta}{2}$ for θ into each of the equations to obtain

$$\cos^2\frac{\theta}{2} = \frac{1 + \cos\theta}{2}$$

$$\sin^2\frac{\theta}{2} = \frac{1 - \cos\theta}{2}$$

Then, for each of the two equations, take the square root of both sides, yielding the following two half-angle identities:

$$\cos\frac{\theta}{2} = \pm\sqrt{\frac{1 + \cos\theta}{2}}$$

$$\sin\frac{\theta}{2} = \pm\sqrt{\frac{1 - \cos\theta}{2}}$$

Be mindful that taking the square roots introduces algebraic signs that are dependent on the quadrant of the involved angle.

In addition, $\tan\frac{\theta}{2} = \frac{\sin\theta}{1 + \cos\theta} = \frac{1 - \cos\theta}{\sin\theta}$, which can be derived from the two formulas above.

EXAMPLE

▶ Use a half-angle identity to find the exact value of $\cos 105°$.

▶ Since the cosine is negative in quadrant II, you have

$$\cos 105° = \cos\left(\frac{210°}{2}\right) = -\sqrt{\frac{1 + \cos 210°}{2}}$$

$$= -\sqrt{\frac{1 + \left(-\frac{\sqrt{3}}{2}\right)}{2}} = -\sqrt{\frac{1 - \frac{\sqrt{3}}{2}}{2}} = -\sqrt{\frac{2 - \sqrt{3}}{4}} = -\frac{1}{2}\sqrt{2 - \sqrt{3}}$$

EXAMPLE

If $\sin\dfrac{\theta}{2} = \dfrac{3}{\sqrt{10}}$ and $\cos\theta$ is negative, find $\tan\theta$ for $0 \le \theta \le 360°$.

Since $\cos\theta$ is negative and $\sin\dfrac{\theta}{2}$ is positive, θ is in quadrant II. Thus,

$\sin^2\dfrac{\theta}{2} = \dfrac{9}{10} = \dfrac{1-\cos\theta}{2}$. Solving yields $\cos\theta = -\dfrac{4}{5}$. It follows that

$\tan\theta = -\dfrac{3}{4}$.

EXAMPLE

Verify that $\tan\dfrac{\theta}{2}\sin\theta = \dfrac{\tan\theta - \sin\theta}{\sin\theta\sec\theta}$ is an identity.

$$\dfrac{\tan\theta - \sin\theta}{\sin\theta\sec\theta} = \dfrac{\sin\theta\left(\dfrac{1}{\cos\theta} - 1\right)}{\sin\theta\sec\theta} = \dfrac{\dfrac{1-\cos\theta}{\cos\theta}}{\dfrac{1}{\cos\theta}}$$

$$= \dfrac{1-\cos\theta}{1} = \dfrac{(1-\cos\theta)(1+\cos\theta)}{1(1+\cos\theta)} = \dfrac{1-\cos^2\theta}{1+\cos\theta}$$

$$= \dfrac{\sin^2\theta}{1+\cos\theta} = \dfrac{\sin\theta}{1+\cos\theta}\sin\theta = \tan\dfrac{\theta}{2}\sin\theta$$

EXERCISE 7-9

For questions 1 to 7, use a half-angle identity to find the exact value.

1. $\sin\dfrac{\pi}{12}$

2. $\tan 15°$

3. $\cos\dfrac{7\pi}{12}$

4. $\cos 15°$

5. $\tan 30°$

6. $\tan 157.5°$

7. $\cos 67.5°$

For questions 8 to 10, verify the identity.

8. $\sin^2\dfrac{\theta}{2} = \dfrac{\sec\theta - 1}{2\sec\theta}$

9. $\dfrac{1+\cos 2\theta}{1-\cos 2\theta} = \cot^2\theta$

10. $\sin\theta\tan\dfrac{\theta}{2} = \dfrac{\tan\theta - \sin\theta}{\sin\theta\sec\theta}$

Sum-to-Product Identities

Following are the sum-to-product identities:

▶ $\cos\theta + \cos\varphi = 2\cos\left(\dfrac{\theta+\varphi}{2}\right)\cos\left(\dfrac{\theta-\varphi}{2}\right)$

▶ $\cos\theta - \cos\varphi = -2\sin\left(\dfrac{\theta+\varphi}{2}\right)\sin\left(\dfrac{\theta-\varphi}{2}\right)$

▶ $\sin\theta + \sin\varphi = 2\sin\left(\dfrac{\theta+\varphi}{2}\right)\cos\left(\dfrac{\theta-\varphi}{2}\right)$

▶ $\sin\theta - \sin\varphi = 2\cos\left(\dfrac{\theta+\varphi}{2}\right)\sin\left(\dfrac{\theta-\varphi}{2}\right)$

EXAMPLE

▶ Express cos 200° + cos 100° as a product of two functions. Leave in terms of sine and/or cosine.

$$\cos 200° + \cos 100° = 2\cos\left(\frac{200° + 100°}{2}\right)\cos\left(\frac{200° - 100°}{2}\right) = 2\cos 150° \cos 50°$$

EXERCISE 7-10

For questions 1 to 7, write the sum or difference as a product.

1. cos 3θ − cos 8θ

2. $\sin\dfrac{3}{4}\theta - \sin\dfrac{1}{4}\theta$

3. cos 4θ + cos 9θ

4. sin 6θ + sin 4θ

5. cos θ + cos 5θ

6. sin 18° − sin 6°. Leave in terms of sine and/or cosine.

7. cos 6θ + cos 2θ

For questions 8 to 10, verify the identity.

8. $\dfrac{\sin\theta + \sin\varphi}{\cos\theta + \cos\varphi} = \tan\dfrac{1}{2}(\theta + \varphi)$

9. $\dfrac{\sin 9\theta - \sin 5\theta}{\sin 14\theta} = \dfrac{\sin 2\theta}{\sin 7\theta}$

10. $\dfrac{\sin 6\theta - \sin 4\theta}{\cos 6\theta + \cos 4\theta} = \tan\theta$

Product-to-Sum Identities

Following are the product-to-sum identities:

$$\blacktriangleright \quad \cos\theta\cos\varphi = \frac{1}{2}\left[\cos(\theta + \varphi) + \cos(\theta - \varphi)\right]$$

$$\blacktriangleright \quad \sin\theta\sin\varphi = \frac{1}{2}\left[\cos(\theta - \varphi) - \cos(\theta + \varphi)\right]$$

$$\blacktriangleright \quad \sin\theta\cos\varphi = \frac{1}{2}\left[\sin(\theta + \varphi) + \sin(\theta - \varphi)\right]$$

$$\blacktriangleright \quad \cos\theta\sin\varphi = \frac{1}{2}\left[\sin(\theta + \varphi) - \sin(\theta - \varphi)\right]$$

EXAMPLE

$\blacktriangleright$ Using a sum or difference of two functions, find the exact value of $\cos 45°\cos 15°$.

$$\cos 45°\cos 15° = \frac{1}{2}[\cos(45° + 15°) + \cos(45° - 15°)]$$

$$= \frac{1}{2}[\cos(60°) + \cos(30°)] = \frac{1}{2}\left[\frac{1}{2} + \frac{\sqrt{3}}{2}\right] = \frac{1 + \sqrt{3}}{4}$$

EXERCISE 7-11

For questions 1 to 6, write the product as a sum or difference.

1. $\sin 3\theta \cos \theta$

2. $\cos 2\theta \cos 6\theta$

3. $\cos 6\theta \sin 4\theta$

4. $\sin 8\theta \sin 4\theta$

5. $\sin 5\theta \cos 8\theta$

6. $\sin 3\theta \cos 7\theta$

7. Using a sum or difference of two functions, find the exact value of $\sin 45° \sin 15°$.

For questions 8 to 10, verify the identity.

8. $\cos 7\theta + \cos 5\theta + 2\cos\theta\cos 2\theta = 4\cos 4\theta\cos 2\theta\cos\theta$

9. $\cos(\theta + 45°)\cos\theta = \frac{\sqrt{2}}{2}\cos\theta[\cos\theta - \sin\theta]$

10. $\sin(\theta + 45°)\sin\theta = \frac{\sqrt{2}}{4}[1 - \cos 2\theta + \sin 2\theta]$

CHAPTER 8

Trigonometric Functions of Real Numbers

Definitions and Basic Concepts of Trigonometric Functions of Real Numbers

In previous lessons, the argument of a trigonometric function was an angle (often denoted by θ). To define a trigonometric function of a real number x, rather than an angle θ, let x equal the real number that corresponds to the radian measure of θ.

The domain of the functions $y = \sin x$ and $y = \cos x$ is the set of all real numbers. The domain of the function $y = \tan x$ is all real numbers *except* the real numbers $\dfrac{\pi}{2} + n\pi$, where n is any integer. The domain of the function $y = \sec x$ is all real numbers *except* the real numbers $\dfrac{\pi}{2} + n\pi$, where n is any integer. The domain of the function $y = \csc x$ is all real numbers *except* the real numbers $n\pi$, where n is any integer. And the domain of the function $y = \cot x$ is all real numbers *except* the real numbers $n\pi$, where n is any integer.

EXAMPLE

▶ When $x = \dfrac{\pi}{4}$, $y = \sin x = \sin\dfrac{\pi}{4} \approx 0.707$.

▶ When $x = 1$, $y = \cos x = \cos 1 \approx 0.540$.

▶ When $x = -1.5$, $y = \tan x = \tan(-1.5) \approx -14.1$.

Set your calculator to radian mode when you are evaluating trigonometric functions of real numbers.

To the reader: Trigonometric functions have now been defined as functions of angles and as functions of real numbers. In your remaining work in trigonometry, be prepared to draw on your understanding of both perspectives.

EXERCISE 8-1

For questions 1 to 5, evaluate $y = \sin x$ for the given value of the real number x. (Round answers to three decimal places, as needed.)

1. 1

2. 20

3. $\dfrac{5\pi}{3}$

4. -1000

5. $-\dfrac{3\pi}{4}$

For questions 6 to 10, evaluate $y = \cos x$ for the given value of the real number x. (Round answers to three decimal places, as needed.)

6. 1

7. 20

8. $\dfrac{5\pi}{3}$

9. -1000

10. $-\dfrac{3\pi}{4}$

For questions 11 to 15, evaluate $y = \tan x$ for the given value of the real number x. (Round answers to three decimal places, as needed.)

11. 1

12. 20

13. $\dfrac{5\pi}{3}$

14. -1000

15. $-\dfrac{3\pi}{4}$

For questions 16 to 20, evaluate $y = \sec x$ for the given value of the real number x. (Round answers to three decimal places, as needed.)

16. 1

17. 20

18. $\dfrac{5\pi}{3}$

19. -1000

20. $-\dfrac{3\pi}{4}$

For questions 21 to 25, evaluate $y = \csc x$ for the given value of the real number x. (Round answers to three decimal places, as needed.)

21. 1

22. 20

23. $\dfrac{5\pi}{3}$

24. -1000

25. $-\dfrac{3\pi}{4}$

For questions 26 to 30, evaluate $y = \cot x$ for the given value of the real number x. (Round answers to three decimal places, as needed.)

26. 1

27. 20

28. $\dfrac{5\pi}{3}$

29. -1000

30. $-\dfrac{3\pi}{4}$

Periodic Functions

The trigonometric functions are distinctive types of functions because they are periodic. Periodic functions repeat their values in regular intervals or periods. Specifically, a **periodic function** is a function f in which there is a positive constant P such that for all x in the domain of f, $f(x + nP) = f(P)$, where n is an integer. The least number p for which this is true is the **period** of f.

EXAMPLE

▶ The functions $y = \sin x, y = \cos x, y = \sec x$, and $y = \csc x$ are periodic functions, each with period 2π.

▶ The functions $y = \tan x$ and $y = \cot x$ are periodic functions, each with period π.

The graph of a periodic function displays a repetitive, cyclical pattern that coincides with the period of the function. Each **cycle** is one repetition of the periodic pattern and has a horizontal length equal to the function's period.

The vertical height of the graph is its **amplitude**. It equals one-half the absolute difference between the function's maximum value and its minimum value; that is, amplitude $= \dfrac{1}{2}\left|\text{maximum value} - \text{minimum value}\right|$.

The **equation of the midline** about which the graph cycles is the equation of the horizontal line halfway between the maximum and the minimum values; thus, the equation of the midline is $y = \dfrac{\text{maximum value} + \text{minimum value}}{2}$.

Not all periodic functions have an amplitude and a midline. For example, the tangent does not have an amplitude (or a midline) because it does not have a maximum or minimum value. Its amplitude is undefined.

The periodic function shown below has period 4, amplitude 1.5, and midline equation $y = 0(x\text{-axis})$.

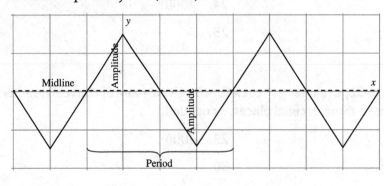

EXERCISE 8-2

For questions 1 and 2, fill in the blank(s) to make a true statement.

1. A periodic function is a function f in which there is a positive constant P such that _____ $= f(P)$, for all x in the domain of f.

2. The amplitude of a periodic function equals one-half the absolute difference between the function's _____ value and its _____ value.

For questions 3 to 5, answer the question for the graph of the periodic function shown below.

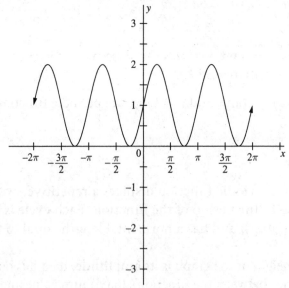

3. What is the function's period?

    a. $\dfrac{\pi}{2}$

    b. π

    c. 2π

    d. 4π

4. What is the function's amplitude?

    a. $\dfrac{1}{2}$

    b. 1

    c. 2

    d. none

5. What is the equation of the function's midline?

    a. $y = 0$

    b. $y = 1$

    c. $y = 2$

    d. none

For questions 6 to 8, answer the question for the graph of the periodic function shown below.

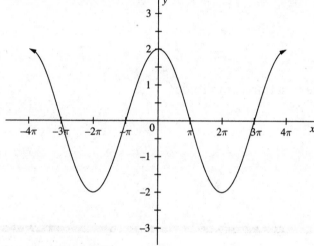

6. What is the function's period?

    a. π

    b. 2π

    b. 3π

    d. 4π

7. What is the function's amplitude?

    a. 1

    b. 2

    c. 4

    d. none

8. What is the equation of the function's midline?

    a. $y = -2$

    b. $y = 0$

    c. $y = 2$

    d. none

For questions 9 and 10, answer the question for the graph of the periodic function shown below.

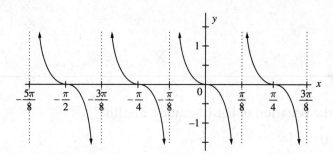

9. What is the function's period?

    a. $\dfrac{\pi}{8}$

    b. $\dfrac{\pi}{4}$

    c. $\dfrac{\pi}{2}$

    d. π

10. What is the function's amplitude?

    a. $\dfrac{1}{2}$

    b. 1

    c. 2

    d. none

CHAPTER 9

Graphs of the Sine Function

The Graph of $y = \sin x$

To graph the function $y = \sin x$, create an x-y table using some familiar values of x in the interval from 0 to 2π. Plot the table's ordered pairs, and then connect the points with a smooth curve. Hint: To make the graph easier to read, stretch the scale on the y-axis.

x	0	$\dfrac{\pi}{6}$	$\dfrac{\pi}{4}$	$\dfrac{\pi}{2}$	$\dfrac{3\pi}{4}$	$\dfrac{5\pi}{6}$	π
$\sin x$	0	$\dfrac{1}{2}$	$\dfrac{\sqrt{2}}{2}$	1	$\dfrac{\sqrt{2}}{2}$	$\dfrac{1}{2}$	0
$x \sin x$	$(0,0)$	$\left(\dfrac{\pi}{6},\dfrac{1}{2}\right)$	$\left(\dfrac{\pi}{4},\dfrac{\sqrt{2}}{2}\right)$	$\left(\dfrac{\pi}{2},1\right)$	$\left(\dfrac{3\pi}{4},\dfrac{\sqrt{2}}{2}\right)$	$\left(\dfrac{5\pi}{6},\dfrac{1}{2}\right)$	$(\pi,0)$

x	$\dfrac{7\pi}{6}$	$\dfrac{5\pi}{4}$	$\dfrac{3\pi}{2}$	$\dfrac{7\pi}{4}$	$\dfrac{11\pi}{6}$	2π
$\sin x$	$-\dfrac{1}{2}$	$-\dfrac{\sqrt{2}}{2}$	-1	$-\dfrac{\sqrt{2}}{2}$	$-\dfrac{1}{2}$	0
$(x,\sin x)$	$\left(\dfrac{7\pi}{6},-\dfrac{1}{2}\right)$	$\left(\dfrac{5\pi}{4},-\dfrac{\sqrt{2}}{2}\right)$	$\left(\dfrac{3\pi}{2},-1\right)$	$\left(\dfrac{7\pi}{4},-\dfrac{\sqrt{2}}{2}\right)$	$\left(\dfrac{11\pi}{6},-\dfrac{1}{2}\right)$	$(2\pi,0)$

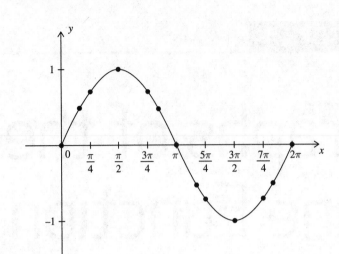

This graph represents only a portion of $y = \sin x$. This portion is the **basic cycle** of $y = \sin x$. The entire graph repeats the pattern of the basic cycle joined end to end in both the positive and negative directions on the x-axis. Below is a graph of $y = \sin x$ that shows several cycles, with the basic cycle plotted with a thicker line style.

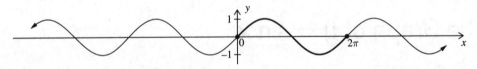

| The sine function is odd, meaning $\sin(-x) = -\sin x$. Graphically, an odd function is symmetric about the origin, as exhibited in the graph of $y = \sin x$. |

The curve is smooth and continuous with a wavelike appearance. It has no vertical asymptotes, holes, gaps, jumps, or corners. (An **asymptote** is a line that the graph of a function gets closer and closer to in at least one direction along the line.) Functions whose graphs have the same wavelike shape of sine functions are **sinusoidal**.

EXERCISE 9-1

Answer the questions using the graph of the function $y = \sin x$ shown below.

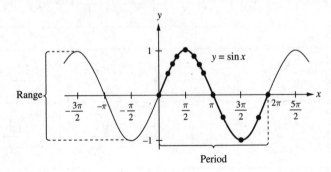

1. What is the range of $y = \sin x$?

2. What is the period of $y = \sin x$?

3. What is the maximum value of $y = \sin x$?

4. What is the minimum value of $y = \sin x$?

5. What is the amplitude of $y = \sin x$?

6. What is the equation of the midline of $y = \sin x$?

7. For what values of x in the interval from $0 \leq x \leq 2\pi$ is $\sin x = 0$?

8. For what values of x in the interval from $0 \leq x \leq 2\pi$ is $\sin x = 1$?

9. For what values of x in the interval from $0 \leq x \leq 2\pi$ is $\sin x = -1$?

10. For what values of x in the interval from $0 \leq x \leq 2\pi$ is $y = \sin x$ positive?

11. For what values of x in the interval from $0 \leq x \leq 2\pi$ is $y = \sin x$ negative?

12. For what values of x in the interval from $0 \leq x \leq 2\pi$ is $y = \sin x$ increasing?

13. For what values of x in the interval from $0 \leq x \leq 2\pi$ is $y = \sin x$ decreasing?

14. Is the value -3 in the domain of $y = \sin x$? Yes or No?

15. Is the value -3 in the range of $y = \sin x$? Yes or No?

The Graph of $y = A \sin x$

The function $y = A \sin x$ is a transformation of the function $y = \sin x$. The constant A is either a vertical stretch factor or a vertical compression factor, and $|A|$ is the amplitude of the graph of $y = A \sin x$. If $|A| > 1$, the graph is stretched vertically, and if $|A| < 1$, the graph is compressed vertically. If $A < 0$, the graph is reflected over the x-axis. The function $y = A \sin x$ has a maximum value of $|A|$ and a minimum value of $-|A|$. Hence, its range is $-|A| \leq y \leq |A|$. The equation of the midline, the zeros, and the period of $y = A \sin x$ are the same as for $y = \sin x$.

EXAMPLE

The function $y = 2 \sin x$ has period 2π, amplitude 2, maximum 2, minimum -2, and midline equation $y = 0$ (see the figure below).

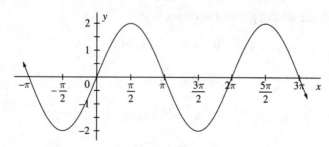

EXAMPLE

The function $y = \dfrac{1}{2}\sin x$ has period 2π, amplitude $\dfrac{1}{2}$, maximum $\dfrac{1}{2}$, minimum $-\dfrac{1}{2}$, and midline equation $y = 0$ (see the figure below).

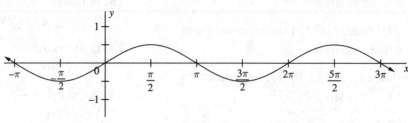

EXAMPLE

The function $y = -\sin x$ is a reflection over the x-axis. It has period 2π, amplitude 1, maximum 1, minimum -1, and midline equation $y = 0$ (see the figure below).

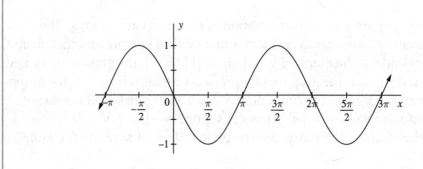

EXERCISE 9-2

For questions 1 to 10, state the amplitude of the given function.

1. $y = 4\sin x$

2. $y = \dfrac{1}{3}\sin x$

3. $y = 1.5\sin x$

4. $y = -5\sin x$

5. $y = \sqrt{2}\sin x$

6. $y = -0.4\sin x$

7. $y = \dfrac{4}{5}\sin x$

8. $y = -10\sin x$

9. $y = 0.75\sin x$

10. $y = -\sqrt{3}\sin x$

For questions 11 to 15, state the range of the given function.

11. $y = 4\sin x$

12. $y = 1.5\sin x$

13. $y = -0.4\sin x$

14. $y = 0.75\sin x$

15. $y = -\sqrt{3}\sin x$

For questions 16 to 20, answer as indicated.

16. What is the maximum value of $y = -10\sin x$?

17. What is the minimum value of $y = \frac{4}{5}\sin x$?

18. For what values of x in the interval $0 \le x \le 2\pi$ is $1.5\sin x = 0$?

19. For what values of x in the interval $0 \le x \le 2\pi$ is $y = -\sqrt{3}\sin x$ a maximum value?

20. For what values of x in the interval $0 \le x \le 2\pi$ is $y = -5\sin x$ a minimum value?

For questions 21 to 25, sketch at least one cycle of the graph of the given function.

21. $y = 4\sin x$

22. $y = 1.5\sin x$

23. $y = -5\sin x$

24. $y = \frac{4}{5}\sin x$

25. $y = -\sqrt{3}\sin x$

The Graph of $y = A \sin Bx$

The function $y = A \sin Bx$ is a transformation of the function $y = \sin x$. The two constants A and B impact the graph of $y = A \sin Bx$ in different ways. The constant A in $y = A \sin Bx$ has the same impact as it does in $y = A \sin x$ (see the previous lesson). It is a vertical stretch (or compression) factor, which affects the graph's amplitude. In contrast, the constant B is either a horizontal stretch factor or a horizontal compression factor, each of which alters

the period. Specifically, the period of $y = A \sin Bx$ is $\dfrac{2\pi}{|B|}$.

If $|B| > 1$, the graph is compressed horizontally so that it completes a full cycle in a shorter period $(< 2\pi)$, and if $|B| < 1$, the graph is stretched horizontally so that it completes a full cycle in a longer period $(> 2\pi)$. If $B < 0$, use your knowledge of odd functions to rewrite $y = A\sin Bx$ as the equivalent function $y = -A\sin|B|x$.

> The absolute value of B is used in $\frac{2\pi}{|B|}$ to ensure that the period is stated as a positive number.

EXAMPLE

The function $y = 3\sin 2x$ has period $\pi\left(= \dfrac{2\pi}{2}\right)$, amplitude 3, maximum 3, minimum -3, and midline equation $y = 0$ (see the figure below).

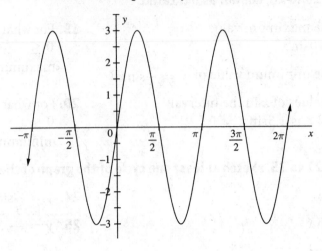

EXAMPLE

The function $y = 3\sin\dfrac{x}{2}$ has period $4\pi\left(= \dfrac{2\pi}{1/2}\right)$, amplitude 3, maximum 3, minimum -3, and midline equation $y = 0$ (see the figure below).

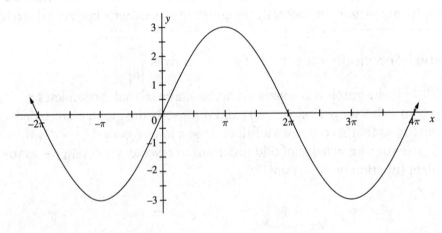

EXAMPLE

▶ The function $y = \sin(-x) = -\sin x$ is a reflection over the x-axis. It has period 2π, amplitude 1, maximum 1, minimum -1, and midline equation $y = 0$ (see the figure below).

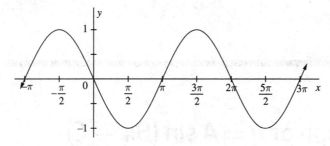

EXERCISE 9-3

For questions 1 to 10, state the period of the given function.

1. $y = \sin 4x$

2. $y = \dfrac{1}{3}\sin 2x$

3. $y = 1.5\sin x$

4. $y = -5\sin 6x$

5. $y = \sqrt{2}\sin\dfrac{x}{3}$

6. $y = -0.6\sin\dfrac{1}{2}x$

7. $y = \dfrac{4}{5}\sin\pi x$

8. $y = -1\sin 2.5x$

9. $y = \dfrac{3}{5}\sin(-8x)$

10. $y = -\sqrt{3}\sin 0.25x$

For questions 11 to 15, answer as indicated.

11. For what values of x in the interval $0 \le x \le 2\pi$ is $\sin 4x = 0$?

12. For what values of x in the interval $0 \le x \le 2\pi$ is $y = \sin 4x$ a maximum?

13. For what values of x in the interval $0 \le x \le 2\pi$ is $y = \sin 4x$ a minimum?

14. For what values of x in the interval $0 \le x \le 2\pi$ is $-0.6\sin\dfrac{1}{2}x = 0$?

15. For what value of x in the interval $0 \le x \le 2\pi$ is $y = -0.6\sin\dfrac{1}{2}x$ a maximum?

For questions 16 to 20, rewrite the function as an equivalent sine function in which the coefficient of x is positive.

16. $y = 3\sin(-2x)$

17. $y = -\sin(-3x)$

18. $y = -4\sin(-2.5x)$

19. $y = \sqrt{2}\sin\left(-\dfrac{x}{3}\right)$

20. $y = \dfrac{4}{5}\sin(-\pi x)$

For questions 21 to 25, sketch at least one cycle of the graph of the given function.

21. $y = -0.6 \sin \dfrac{1}{2} x$

22. $y = \sin 4x$

23. $y = -1 \sin 2.5x$

24. $y = \sqrt{2} \sin \dfrac{x}{3}$

25. $y = \dfrac{4}{5} \sin \pi x$

The Graph of $y = A \sin (Bx - C)$

The graph of the function $y = A\sin(Bx - C) = A\sin B\left(x - \dfrac{C}{B}\right)$ coincides

with the graph of $y = A\sin Bx$ shifted horizontally by $\dfrac{C}{B}$ units. The shift

is to the right when $\dfrac{C}{B} > 0$, and to the left when $\dfrac{C}{B} < 0$. The number $\dfrac{C}{B}$ is

the **phase shift**. In identifying the phase shift, you will find it helpful to write

$y = A\sin(Bx - C)$ in **shift form** as $y = A\sin B\left(x - \dfrac{C}{B}\right)$. Solve the equations

$x - \dfrac{C}{B} = 0$ and $x - \dfrac{C}{B} = \dfrac{2\pi}{|B|}$ to determine the left and right endpoints,

respectively, of an interval that corresponds to one cycle of the graph.

EXAMPLE

The function $y = 3\sin\left(x + \dfrac{\pi}{6}\right) = 3\sin\left(x - \left(-\dfrac{\pi}{6}\right)\right)$ has period

2π, amplitude 3, and a phase shift of $\dfrac{\pi}{6}$ units to the left. Solving

$x - \left(-\dfrac{\pi}{6}\right) = 0$ and $x - \left(-\dfrac{\pi}{6}\right) = 2\pi$ yields $\left[-\dfrac{\pi}{6}, \dfrac{11\pi}{6}\right]$ as a one-cycle

interval of the graph (see the figure below in which the portion of the graph

corresponding to the interval $\left[-\dfrac{\pi}{6}, \dfrac{11\pi}{6}\right]$ is plotted in a thicker line style).

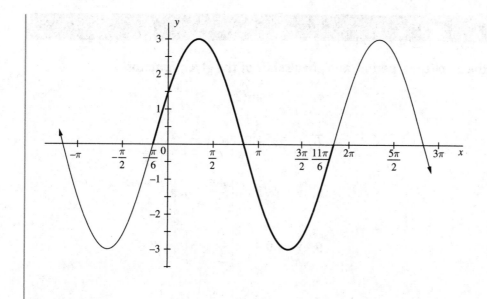

The function $y = 2\sin\left(4x - \dfrac{4\pi}{3}\right) = 2\sin\left(4\left(x - \dfrac{\pi}{3}\right)\right)$ has period

$\dfrac{\pi}{2}\left(= \dfrac{2\pi}{4}\right)$, amplitude 2, and a phase shift of $\dfrac{\pi}{3}$ units to the right. Solving

$x - \dfrac{\pi}{3} = 0$ and $x - \dfrac{\pi}{3} = \dfrac{\pi}{2}$ yields $\left[\dfrac{\pi}{3}, \dfrac{5\pi}{6}\right]$ as a one-cycle interval of the

graph (see the figure below in which the portion of the graph corresponding

to the interval $\left[\dfrac{\pi}{3}, \dfrac{5\pi}{6}\right]$ is plotted in a thicker line style).

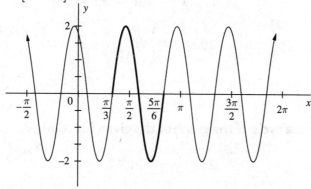

EXERCISE 9-4

For questions 1 to 10, state the amplitude, period, and phase shift of the given function.

1. $y = 5\sin\left(x - \dfrac{\pi}{4}\right)$

2. $y = 5\sin\left(x + \dfrac{\pi}{4}\right)$

3. $y = \sin\left(2x + \dfrac{\pi}{3}\right)$

4. $y = \dfrac{1}{2}\sin(3x - \pi)$

5. $y = -3\sin\left(3x - \dfrac{\pi}{4}\right)$

6. $y = 6\sin(2x - 3)$

7. $y = -\sin\left(\pi x - \dfrac{\pi}{3}\right)$

8. $y = \sqrt{2}\sin\left(\dfrac{\pi x}{2} + \pi\right)$

9. $y = 0.4\sin\left(\dfrac{x}{2} + 3\right)$

10. $y = -2\sin\left(\dfrac{2x}{3} - 1\right)$

For questions 11 to 15, rewrite the function as an equivalent sine function in which the coefficient of x is positive.

11. $y = -3\sin\left(-\dfrac{2x}{3} + \dfrac{\pi}{4}\right)$

12. $y = -2\sin\left(-2x + \dfrac{\pi}{2}\right)$

13. $y = \dfrac{1}{2}\sin(-2x + 4)$

14. $y = 3\sin(-2\pi x + \pi)$

15. $y = \dfrac{3}{4}\sin\left(-x - \dfrac{\pi}{3}\right)$

For questions 16 to 20, use the equations $x - \dfrac{C}{B} = 0$ and $x - \dfrac{C}{B} = \dfrac{2\pi}{|B|}$ to determine a one-cycle interval.

16. $y = -3\sin\left(-\dfrac{2x}{3} + \dfrac{\pi}{4}\right)$

17. $y = -2\sin\left(-2x + \dfrac{\pi}{2}\right)$

18. $y = \dfrac{1}{2}\sin(-2x + 4)$

19. $y = 3\sin(-2\pi x + \pi)$

20. $y = \dfrac{3}{4}\sin\left(-x - \dfrac{\pi}{3}\right)$

For questions 21 to 25, sketch at least one cycle of the graph of the given function.

21. $y = -3\sin\left(-\dfrac{2x}{3} + \dfrac{\pi}{4}\right)$

22. $y = -2\sin\left(-2x + \dfrac{\pi}{2}\right)$

23. $y = \dfrac{1}{2}\sin(-2x + 4)$

24. $y = 3\sin(-2\pi x + \pi)$

25. $y = \dfrac{3}{4}\sin\left(-x - \dfrac{\pi}{3}\right)$

The Graph of $y = A \sin(Bx - C) + D$

The graph of the function $y = A\sin(Bx - C) + D = A\sin B\left(x - \dfrac{C}{B}\right) + D$

coincides with the graph of $y = A\sin(Bx - C)$ shifted vertically by D units. When $D > 0$, the shift is $|D|$ units upward, and when $D < 0$, the shift is $|D|$ units downward. Instead of $y = 0$, the midline for $y = A\sin(Bx - C) + D$ is $y = D$.

EXAMPLE

▶ The function $y = 3\sin\left(x + \dfrac{\pi}{6}\right) + 2 = 3\sin\left(x - \left(-\dfrac{\pi}{6}\right)\right) + 2$ has

period 2π, amplitude 3, a phase shift of $\dfrac{\pi}{6}$ units to the left, a vertical shift

of 2 units upward, and a midline equation of $y = 2$. Solving

$x - \left(-\dfrac{\pi}{6}\right) = 0$ and $x - \left(-\dfrac{\pi}{6}\right) = 2\pi$ yields $\left[-\dfrac{\pi}{6}, \dfrac{11\pi}{6}\right]$ as a one-cycle

interval of the graph. See the following graph of

$y = 3\sin\left(x + \dfrac{\pi}{6}\right) + 2 = 3\sin\left(x - \left(-\dfrac{\pi}{6}\right)\right) + 2$ in which the midline $y = 2$ is

plotted in grayscale as a dashed line and the portion of the graph

corresponding to the interval $\left[-\dfrac{\pi}{6}, \dfrac{11\pi}{6}\right]$ is plotted in a thicker line style.

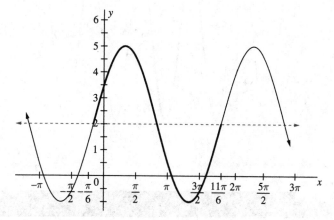

EXAMPLE

The function $y = 2\sin\left(4x - \dfrac{4\pi}{3}\right) - 3 = 2\sin\left(4\left(x - \dfrac{\pi}{3}\right)\right) - 3$

has period $\dfrac{\pi}{2}\left(= \dfrac{2\pi}{4}\right)$, amplitude 2, a phase shift of $\dfrac{\pi}{3}$ units to the right, a vertical shift of 3 units downward, and a midline equation of $y = -3$. Solving $x - \dfrac{\pi}{3} = 0$ and $x - \dfrac{\pi}{3} = \dfrac{\pi}{2}$ yields $\left[\dfrac{\pi}{3}, \dfrac{5\pi}{6}\right]$ as a one-cycle interval of the graph. See the following graph of $y = 2\sin\left(4x - \dfrac{4\pi}{3}\right) - 3 = 2\sin\left(4\left(x - \dfrac{\pi}{3}\right)\right) - 3$ in which the midline $y = -3$ is plotted in grayscale as a dashed line and the portion of the graph corresponding to the interval $\left[\dfrac{\pi}{3}, \dfrac{5\pi}{6}\right]$ is plotted in a thicker line style.

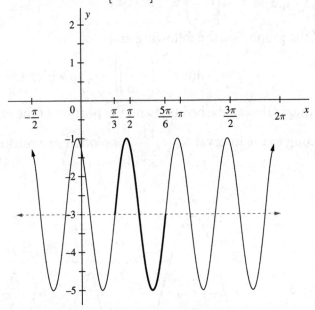

EXERCISE 9-5

For questions 1 to 10, describe the horizontal and vertical shifts of the given function.

1. $y = 5\sin\left(x - \dfrac{\pi}{4}\right) - 1$

2. $y = 5\sin\left(x + \dfrac{\pi}{4}\right) + 6$

3. $y = \sin\left(2x + \dfrac{\pi}{3}\right) - \dfrac{1}{2}$

4. $y = \dfrac{1}{2}\sin(3x - \pi) + \sqrt{3}$

5. $y = -3\sin\left(3x - \dfrac{\pi}{4}\right) - 2.5$

6. $y = 6\sin(2x - 3) + \dfrac{5}{4}$

7. $y = -\sin\left(\pi x - \dfrac{\pi}{3}\right) + 21$

8. $y = \sqrt{2}\sin\left(\dfrac{\pi x}{2} + \pi\right) - 5.6$

9. $y = 0.4\sin\left(\dfrac{x}{2} + 3\right) - 9$

10. $y = -2\sin\left(\dfrac{2x}{3} - 1\right) + 45$

For questions 11 to 15, write the equation of the midline of the given function.

11. $y = 6\sin(2x - 3) + \dfrac{5}{4}$

12. $y = -\sin\left(\pi x - \dfrac{\pi}{3}\right) + 21$

13. $y = \sqrt{2}\sin\left(\dfrac{\pi x}{2} + \pi\right) - 5.6$

14. $y = 0.4\sin\left(\dfrac{x}{2} + 3\right) - 9$

15. $y = -2\sin\left(\dfrac{2x}{3} - 1\right) + 45$

For questions 16 to 19, sketch at least one cycle of the graph of the given function.

16. $y = 5\sin\left(x - \dfrac{\pi}{4}\right) - 1$

17. $y = 2\sin\left(x + \dfrac{\pi}{4}\right) + 3$

18. $y = -\sin\left(-2x + \dfrac{\pi}{3}\right) + \dfrac{1}{2}$

19. $y = \dfrac{1}{2}\sin(-3x - \pi) + 3$

20. The current I (in amperes) in a wire of an alternating current circuit is given by the function $I(t) = 15\sin 120\pi t$, where t is the elapsed time in seconds. What is the period of $I(t)$? What is the maximum value of the current?

CHAPTER 10

Graphs of the Cosine Function

The Graph of $y = \cos x$

To graph the function $y = \cos x$, create an x-y table using some familiar values of x in the interval from 0 to 2π. Plot the table's ordered pairs, and then connect the points with a smooth curve.

x	0	$\dfrac{\pi}{4}$	$\dfrac{\pi}{3}$	$\dfrac{\pi}{2}$	$\dfrac{2\pi}{3}$	$\dfrac{3\pi}{4}$
$\cos x$	1	$\dfrac{\sqrt{2}}{2}$	$\dfrac{1}{2}$	0	$-\dfrac{1}{2}$	$-\dfrac{\sqrt{2}}{2}$
$(x, \cos x)$	$(0, 1)$	$\left(\dfrac{\pi}{4}, \dfrac{\sqrt{2}}{2}\right)$	$\left(\dfrac{\pi}{3}, \dfrac{1}{2}\right)$	$\left(\dfrac{\pi}{2}, 0\right)$	$\left(\dfrac{2\pi}{3}, -\dfrac{1}{2}\right)$	$\left(\dfrac{3\pi}{4}, -\dfrac{\sqrt{2}}{2}\right)$

x	π	$\dfrac{5\pi}{4}$	$\dfrac{4\pi}{3}$	$\dfrac{3\pi}{2}$	$\dfrac{5\pi}{3}$	$\dfrac{7\pi}{4}$	2π
$\cos x$	-1	$-\dfrac{\sqrt{2}}{2}$	$-\dfrac{1}{2}$	0	$\dfrac{1}{2}$	$\dfrac{\sqrt{2}}{2}$	1
$(x, \cos x)$	$(\pi, -1)$	$\left(\dfrac{5\pi}{4}, -\dfrac{\sqrt{2}}{2}\right)$	$\left(\dfrac{4\pi}{3}, -\dfrac{1}{2}\right)$	$\left(\dfrac{3\pi}{2}, 0\right)$	$\left(\dfrac{5\pi}{3}, \dfrac{1}{2}\right)$	$\left(\dfrac{7\pi}{4}, \dfrac{\sqrt{2}}{2}\right)$	$(2\pi, 1)$

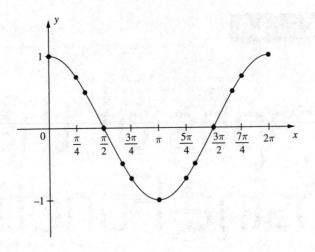

This graph represents only a portion of $y = \cos x$. This portion is the **basic cycle** of $y = \cos x$. The entire graph repeats the pattern of the basic cycle joined end to end in both the positive and negative directions on the x-axis. Below is a graph of $y = \cos x$ that shows several cycles, with the basic cycle plotted with a thicker line style.

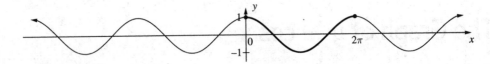

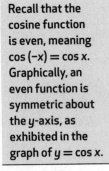

Recall that the cosine function is even, meaning $\cos(-x) = \cos x$. Graphically, an even function is symmetric about the y-axis, as exhibited in the graph of $y = \cos x$.

Like the sine function, the curve is smooth, continuous, and sinusoidal with no vertical asymptotes, holes, gaps, jumps, or corners.

EXERCISE 10-1

Answer the following questions using the graph of the function $y = \cos x$ shown below.

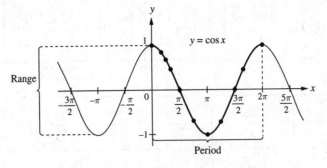

1. What is the range of $y = \cos x$?

2. What is the period of $y = \cos x$?

3. What is the maximum value of $y = \cos x$?

4. What is the minimum value of $y = \cos x$?

5. What is the amplitude of $y = \cos x$?

6. What is the equation of the midline of $y = \cos x$?

7. For what values of x in the interval from $0 \leq x \leq 2\pi$ is $\cos x = 0$?

8. For what values of x in the interval from $0 \leq x \leq 2\pi$ is $\cos x = 1$?

9. For what values of x in the interval from $0 \leq x \leq 2\pi$ is $\cos x = -1$?

10. For what values of x in the interval from $0 \leq x \leq 2\pi$ is $y = \cos x$ positive?

11. For what values of x in the interval from $0 \leq x \leq 2\pi$ is $y = \cos x$ negative?

12. For what values of x in the interval from $0 \leq x \leq 2\pi$ is $y = \cos x$ increasing?

13. For what values of x in the interval from $0 \leq x \leq 2\pi$ is $y = \cos x$ decreasing?

14. Is the value -5 in the domain of $y = \cos x$? Yes or No?

15. Is the value -5 in the range of $y = \cos x$? Yes or No?

The Graph of $y = A \cos (Bx - C) + D$

The function $y = A\cos(Bx - C) + D = A\cos B\left(x - \dfrac{C}{B}\right) + D$ is a

transformation of the function $y = \cos x$. Its graph has an amplitude of $|A|$ and

a period of $\dfrac{2\pi}{|B|}$. The number $\dfrac{C}{B}$ induces a horizontal phase shift. The shift is to

the right when $\dfrac{C}{B} > 0$, and to the left when $\dfrac{C}{B} < 0$. The number D induces a

vertical shift. When $D > 0$, the shift is $|D|$ units upward, and when $D < 0$, the

shift is $|D|$ units downward. The midline of the graph is $y = D$. Note: If $B < 0$,

use your knowledge of even functions to rewrite $y = A \cos (Bx - C) + D$ as the

equivalent function $y = A\cos(|B|x + C) + D$.

The graph of $y = 3\cos\left(x + \dfrac{\pi}{6}\right) + 2 = 3\cos\left(x - \left(-\dfrac{\pi}{6}\right)\right) + 2$

has period 2π, amplitude 3, a phase shift of $\dfrac{\pi}{6}$ units to the left, a

vertical shift of 2 units upward, and a midline equation of $y = 2$.

Solving $x - \left(-\dfrac{\pi}{6}\right) = 0$ and $x - \left(-\dfrac{\pi}{6}\right) = 2\pi$ yields $\left[-\dfrac{\pi}{6}, \dfrac{11\pi}{6}\right]$

as a one-cycle interval of the graph. See the following graph of

$y = 3\cos\left(x + \dfrac{\pi}{6}\right) + 2 = 3\cos\left(x - \left(-\dfrac{\pi}{6}\right)\right) + 2$ in which the midline

$y = 2$ is plotted in grayscale as a dashed line and the portion of the graph

corresponding to the interval $\left[-\dfrac{\pi}{6}, \dfrac{11\pi}{6}\right]$ is plotted in a thicker line style.

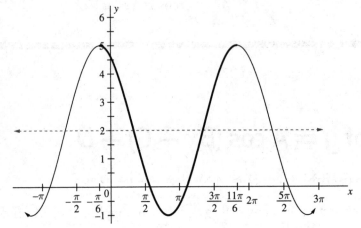

The function $y = 2\cos\left(4x - \dfrac{4\pi}{3}\right) - 3 = 2\cos\left(4\left(x - \dfrac{\pi}{3}\right)\right) - 3$

has period $\dfrac{\pi}{2}\left(= \dfrac{2\pi}{4}\right)$, amplitude 2, a phase shift of $\dfrac{\pi}{3}$ units to the

right, a vertical shift of 3 units downward and a midline equation

of $y = -3$. Solving $x - \dfrac{\pi}{3} = 0$ and $x - \dfrac{\pi}{3} = \dfrac{\pi}{2}$ yields $\left[\dfrac{\pi}{3}, \dfrac{5\pi}{6}\right]$

as a one-cycle interval of the graph. See the following graph of

$y = 2\cos\left(4x - \dfrac{4\pi}{3}\right) - 3 = 2\cos\left(4\left(x - \dfrac{\pi}{3}\right)\right) - 3$ in which the midline

$y = -3$ is plotted in grayscale as a dashed line and the portion of the

graph corresponding to the interval $\left[\dfrac{\pi}{3}, \dfrac{5\pi}{6}\right]$ is plotted in a thicker line style.

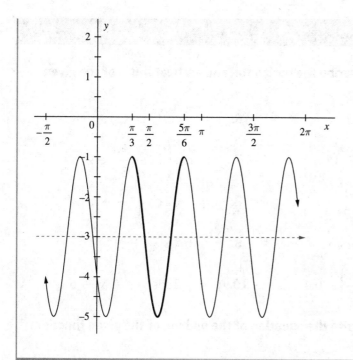

The function $y = \cos\left(x - \dfrac{\pi}{2}\right)$ has period 2π, amplitude 1, a phase shift of

$\dfrac{\pi}{2}$ units to the right, no vertical shift, and a midline equation of

$y = 0$. Solving $x - \dfrac{\pi}{2} = 0$ and $x - \dfrac{\pi}{2} = 2\pi$ yields $\left[\dfrac{\pi}{2}, \dfrac{5\pi}{2}\right]$ as a one-cycle

interval of the graph. See the following graph of $y = \cos\left(x - \dfrac{\pi}{2}\right)$ in

which the interval $\left[\dfrac{\pi}{2}, \dfrac{5\pi}{2}\right]$ is plotted in a thicker line style.

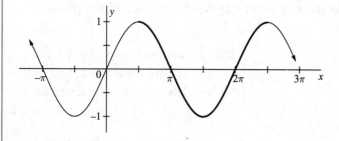

The graph of $y = \cos x$ shifted $\dfrac{\pi}{2}$ units to the right is identical to the

graph of $y = \sin x$. Conversely, the graph of $y = \sin x$ shifted $\dfrac{\pi}{2}$ units to

the left is identical to the graph of $y = \cos x$.

EXERCISE 10-2

For questions 1 to 10, describe the horizontal and vertical shifts of the given function.

1. $y = 3\cos\left(x - \dfrac{\pi}{4}\right) - 2$

2. $y = -5\cos\left(x + \dfrac{\pi}{4}\right) + 1$

3. $y = \cos\left(2x + \dfrac{\pi}{6}\right) + \dfrac{1}{2}$

4. $y = \dfrac{1}{2}\cos(2x - \pi) + \sqrt{5}$

5. $y = -0.6\cos\left(3x - \dfrac{\pi}{2}\right) - 0.4$

6. $y = \dfrac{4}{3}\cos(3x - 2) + \dfrac{5}{4}$

7. $y = \cos\left(\pi x - \dfrac{\pi}{3}\right) + 2$

8. $y = -\sqrt{2}\cos\left(\dfrac{\pi x}{4} + \pi\right) + \sqrt{2}$

9. $y = 6\cos\left(\dfrac{x}{3} + 3\right) - 2$

10. $y = -2\cos\left(\dfrac{4x}{3} - 3\right) + 6$

For questions 11 to 15, write the equation of the midline of the given function.

11. $y = 3\cos\left(x - \dfrac{\pi}{4}\right) - 2$

12. $y = -5\cos\left(x + \dfrac{\pi}{4}\right) + 1$

13. $y = \cos\left(2x + \dfrac{\pi}{6}\right) + \dfrac{1}{2}$

14. $y = \dfrac{1}{2}\cos(2x - \pi) + \sqrt{5}$

15. $y = -0.6\cos\left(3x - \dfrac{\pi}{2}\right) - 0.4$

For questions 16 to 20, rewrite the function as an equivalent cosine function in which the coefficient of *x* is positive.

16. $y = 3\cos(-2x)$

17. $y = -\cos\left(-3x + \dfrac{\pi}{4}\right)$

18. $y = -4\cos\left(-1.5x - \dfrac{2\pi}{3}\right) + 2$

19. $y = \sqrt{2}\cos\left(-\dfrac{x}{3} + \dfrac{\pi}{6}\right)$

20. $y = \dfrac{4}{5}\cos(-\pi x - 4) - \dfrac{3}{2}$

For questions 21 to 24, sketch at least one cycle of the graph of the given function.

21. $y = 5\cos\left(x - \dfrac{\pi}{4}\right) - 1$

22. $y = 2\cos\left(x + \dfrac{\pi}{4}\right) + 3$

23. $y = -\cos\left(-2x + \dfrac{\pi}{3}\right) + \dfrac{1}{2}$

24. $y = \dfrac{1}{2}\cos(-3x - \pi) + 3$

25. The *y*-coordinate of a circular path centered at the origin that models an object traveling at constant angular speed is given by the function $y = 10\cos\left(\pi t - \dfrac{\pi}{12}\right)$, where *t* is the time in seconds. What is the period of the function? What is the maximum value of *y*?

CHAPTER 11

Graphs of the Tangent Function

The Graph of $y = \tan x$

By definition, $\tan x = \dfrac{\sin x}{\cos x}$ (see Chapter 7 for reference). Therefore, the domain of the tangent function is the set of real numbers except those numbers for which $\cos x = 0$. Thus, for the function $y = \tan x$, you specify that $x \neq \dfrac{\pi}{2} + n\pi$, where n is any integer. The graph of $y = \tan x$ has vertical asymptotes at these excluded values of x. The x-intercepts of $y = \tan x$ occur at $x = n\pi$, where n is any integer. The range of $y = \tan x$ is R, the set of real numbers. It has period π and undefined amplitude.

EXAMPLE

Below is a graph of $y = \tan x$ that shows several cycles, with the basic cycle in the interval $\left(-\dfrac{\pi}{2}, \dfrac{\pi}{2}\right)$ plotted with a thicker line style.

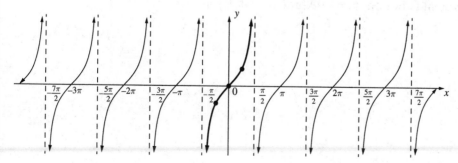

The tangent function is odd, meaning $\tan(-x) = -\tan x$. Graphically, an odd function is symmetric about the origin, as exhibited in the graph of $y = \tan x$.

113

For questions 1 to 7, fill in the blank to make a true statement.

1. By definition, $\tan x =$ _____.

2. The domain of the tangent function excludes numbers for which $\cos x =$ _____.

3. The range of the tangent function is _____.

4. The vertical asymptotes of $y = \tan x$ occur at $x =$ _____.

5. The x-intercepts of $y = \tan x$ occur at $x =$ _____.

6. The period of $y = \tan x$ is _____.

7. The amplitude of $y = \tan x$ is _____.

For questions 8 to 10, indicate whether the statement is True or False.

8. The number $-\dfrac{5\pi}{2}$ is in the domain of the tangent function.

9. The number 0 is in the domain of the tangent function.

10. The number $-\dfrac{5\pi}{2}$ is in the range of the tangent function.

The Graph of $y = A \tan (Bx - C) + D$

The function $y = A \tan(Bx - C) + D$ has domain $x \neq \dfrac{C}{B} + \dfrac{n\pi}{2B}$, where n is an odd integer, and range $(-\infty, \infty)$. Its graph has undefined amplitude, stretching factor of $|A|$, period $\dfrac{\pi}{|B|}$, horizontal phase shift $\dfrac{C}{B}$, vertical shift D, and vertical asymptotes at $x = \dfrac{C}{B} + \dfrac{n\pi}{2B}$, where n is an odd integer. Note: If $B < 0$, use your knowledge of odd functions to rewrite $y = A \tan(Bx - C) + D$ as the equivalent function $y = -A\tan(|B| x + C) + D$.

EXAMPLE

▶ The graph of $y = 2\tan\left(\dfrac{x}{4}\right) - 3$ shown below has period 4π, stretching factor 2, and vertical shift of 3 units downward.

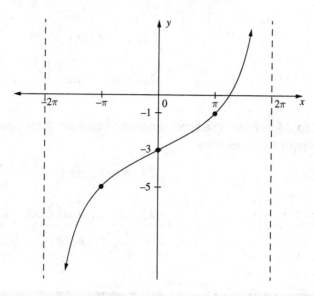

EXERCISE 11-2

For questions 1 to 5, determine the stretching factor and the period of the given function.

1. $y = 3\tan\left(x - \dfrac{\pi}{4}\right) - 2$

2. $y = -5\tan\left(x + \dfrac{\pi}{4}\right) + 1$

3. $y = 4\tan\left(2x + \dfrac{\pi}{6}\right) + \dfrac{1}{2}$

4. $y = \dfrac{1}{2}\tan(2x - \pi) + \sqrt{5}$

5. $y = -0.6\tan\left(3x - \dfrac{\pi}{2}\right) - 0.4$

For questions 6 to 10, describe the horizontal and vertical shifts of the graph of the given function.

6. $y = \dfrac{4}{3}\tan(3x - 2) + \dfrac{5}{4}$

7. $y = 8\tan\left(\pi x - \dfrac{\pi}{3}\right) + 2$

8. $y = -\sqrt{2}\tan\left(\dfrac{\pi x}{4} + \pi\right) + \sqrt{2}$

9. $y = 6\tan\left(\dfrac{x}{3} + 3\right) - 2$

10. $y = -2\tan\left(\dfrac{4x}{3} - 3\right) + 6$

For questions 11 to 20, determine the asymptotes for the graph of the given function.

11. $y = 3\tan\left(x - \dfrac{\pi}{4}\right) - 2$

12. $y = -5\tan\left(x + \dfrac{\pi}{4}\right) + 1$

13. $y = 4\tan\left(2x + \dfrac{\pi}{6}\right) + \dfrac{1}{2}$

14. $y = \dfrac{1}{2}\tan(2x - \pi) + \sqrt{5}$

15. $y = -0.6\tan\left(3x - \dfrac{\pi}{2}\right) - 0.4$

16. $y = \dfrac{4}{3}\tan(3x - 2) + \dfrac{5}{4}$

17. $y = 8\tan\left(\pi x - \dfrac{\pi}{3}\right) + 2$

18. $y = -\sqrt{2}\tan\left(\dfrac{\pi x}{4} + \pi\right) + \sqrt{2}$

19. $y = 6\tan\left(\dfrac{x}{3} + 3\right) - 2$

20. $y = -2\tan\left(\dfrac{4x}{3} - 3\right) + 6$

For questions 21 to 24, rewrite the function as an equivalent tangent function in which the coefficient of x is positive.

21. $y = 3\tan(-2x)$

22. $y = -\tan\left(-3x + \dfrac{\pi}{4}\right)$

23. $y = -4\tan\left(-1.5x - \dfrac{2\pi}{3}\right) + 2$

24. $y = \sqrt{2}\tan\left(-\dfrac{x}{3} + \dfrac{\pi}{6}\right)$

25. Sketch at least one cycle of the graph of $y = -\tan\left(-x + \dfrac{\pi}{3}\right)$.

CHAPTER 12

Graphs of the Secant, Cosecant, and Cotangent Functions

The Graph of $y = A \sec(Bx - C) + D$

The function $y = A\sec(Bx - C) + D$ has domain $x \neq \dfrac{C}{B} + \dfrac{n\pi}{2B}$, where

n is an odd integer, and range $(-\infty, -|A|) \cup (|A|, \infty)$. Its graph has undefined

amplitude, stretching factor of $|A|$, period $\dfrac{2\pi}{|B|}$, horizontal phase shift $\dfrac{C}{B}$,

vertical shift D, and vertical asymptotes at $x = \dfrac{C}{B} + \dfrac{n\pi}{2B}$,

where n is an odd integer. Note: If $B < 0$, use your knowledge of
even functions to rewrite $y = A\sec(Bx - C) + D$ as the equivalent
function $y = A\sec(|B|x + C) + D$.

The graph of $y = \sec(3x - 2\pi) + 4$ shown below has period $\dfrac{2\pi}{3}$, stretching factor 1, a horizontal shift of $\dfrac{2\pi}{3}$ to the right, and a vertical shift of 4 units upward. Also shown is the graph of the reciprocal function $y = \cos(3x - 2\pi) + 4$ as a dashed curve.

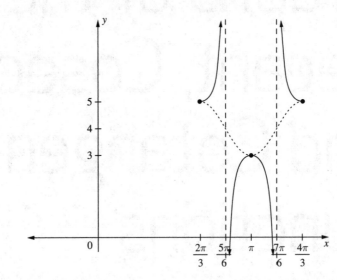

EXERCISE 12-1

For questions 1 to 5, determine the stretching factor and the period of the given function.

1. $y = 3\sec\left(x - \dfrac{\pi}{4}\right) - 2$

2. $y = -5\sec\left(x + \dfrac{\pi}{4}\right) + 1$

3. $y = 4\sec\left(2x + \dfrac{\pi}{6}\right) + \dfrac{1}{2}$

4. $y = \dfrac{1}{2}\sec(2x - \pi) + \sqrt{5}$

5. $y = -0.6\sec\left(3x - \dfrac{\pi}{2}\right) - 0.4$

For questions 6 to 10, describe the horizontal and vertical shifts of the graph of the given function.

6. $y = \dfrac{4}{3}\sec(3x - 2) + \dfrac{5}{4}$

7. $y = 8\sec\left(\pi x - \dfrac{\pi}{3}\right) + 2$

8. $y = -\sqrt{2}\sec\left(\dfrac{\pi x}{4} + \pi\right) + \sqrt{2}$

9. $y = 6\sec\left(\dfrac{x}{3} + 3\right) - 2$

10. $y = -2\sec\left(\dfrac{4x}{3} - 3\right) + 6$

For questions 11 to 20, determine the asymptotes for the graph of the given function.

11. $y = 3\sec\left(x - \dfrac{\pi}{4}\right) - 2$

12. $y = -5\sec\left(x + \dfrac{\pi}{4}\right) + 1$

13. $y = 4\sec\left(2x + \dfrac{\pi}{6}\right) + \dfrac{1}{2}$

14. $y = \dfrac{1}{2}\sec(2x - \pi) + \sqrt{5}$

15. $y = -0.6\sec\left(3x - \dfrac{\pi}{2}\right) - 0.4$

16. $y = \dfrac{4}{3}\sec(3x - 2) + \dfrac{5}{4}$

17. $y = 8\sec\left(\pi x - \dfrac{\pi}{3}\right) + 2$

18. $y = -\sqrt{2}\sec\left(\dfrac{\pi x}{4} + \pi\right) + \sqrt{2}$

19. $y = 6\sec\left(\dfrac{x}{3} + 3\right) - 2$

20. $y = -2\sec\left(\dfrac{4x}{3} - 3\right) + 6$

For questions 21 to 24, rewrite the function as an equivalent secant function in which the coefficient of x is positive.

21. $y = 3\sec(-2x)$

22. $y = -\sec\left(-3x + \dfrac{\pi}{4}\right)$

23. $y = -4\sec\left(-1.5x - \dfrac{2\pi}{3}\right) + 2$

24. $y = \sqrt{2}\sec\left(-\dfrac{x}{3} + \dfrac{\pi}{6}\right)$

25. Sketch at least one cycle of the graph

of $y = \sec\left(x - \dfrac{\pi}{2}\right)$.

The Graph of $y = A\csc(Bx - C) + D$

The function $y = A\csc(Bx - C) + D$ has domain $x \neq \dfrac{C}{B} + \dfrac{n\pi}{B}$, where

n is an integer, and range $(-\infty, -|A|) \cup (|A|, \infty)$. Its graph has undefined

amplitude, stretching factor of $|A|$, period $\dfrac{2\pi}{|B|}$, horizontal phase shift $\dfrac{C}{B}$,

vertical shift D, and vertical asymptotes at $x = \dfrac{C}{B} + \dfrac{n\pi}{B}$, where n is

an integer. Note: If $B < 0$, use your knowledge of odd functions to rewrite

$y = A\csc(Bx - C) + D$ as the equivalent function $y = -A\csc(|B|\, x + C) + D$.

The graph of $y = \csc\left(-x - \dfrac{\pi}{4}\right) - 2 = -\csc\left(x - \left(-\dfrac{\pi}{4}\right)\right) - 2$ shown

below has period π, stretching factor 1, a horizontal shift of $\dfrac{\pi}{4}$ to the left,

and a vertical shift of 2 units downward. Also shown is the graph of the

reciprocal function $y = \sin\left(-x - \dfrac{\pi}{4}\right) - 2$ as a dashed curve.

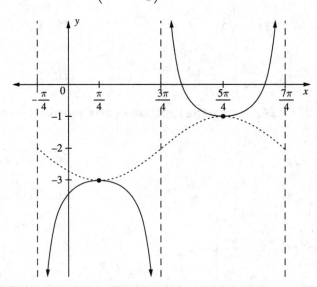

EXERCISE 12-2

For questions 1 to 5, determine the stretching factor and the period of the given function.

1. $y = 3\csc\left(x - \dfrac{\pi}{4}\right) - 2$

2. $y = -5\csc\left(x + \dfrac{\pi}{4}\right) + 1$

3. $y = 4\csc\left(2x + \dfrac{\pi}{6}\right) + \dfrac{1}{2}$

4. $y = \dfrac{1}{2}\csc(2x - \pi) + \sqrt{5}$

5. $y = -0.6\csc\left(3x - \dfrac{\pi}{2}\right) - 0.4$

For questions 6 to 10, describe the horizontal and vertical shifts of the graph of the given function.

6. $y = \dfrac{4}{3}\csc(3x - 2) + \dfrac{5}{4}$

7. $y = 8\csc\left(\pi x - \dfrac{\pi}{3}\right) + 2$

8. $y = -\sqrt{2}\csc\left(\dfrac{\pi x}{4} + \pi\right) + \sqrt{2}$

9. $y = 6\csc\left(\dfrac{x}{3} + 3\right) - 2$

10. $y = -2\csc\left(\dfrac{4x}{3} - 3\right) + 6$

For questions 11 to 20, determine the asymptotes for the graph of the given function.

11. $y = 3\csc\left(x - \dfrac{\pi}{4}\right) - 2$

12. $y = -5\csc\left(x + \dfrac{\pi}{4}\right) + 1$

13. $y = 4\csc\left(2x + \dfrac{\pi}{6}\right) + \dfrac{1}{2}$

14. $y = \dfrac{1}{2}\csc(2x - \pi) + \sqrt{5}$

15. $y = -0.6\csc\left(3x - \dfrac{\pi}{2}\right) - 0.4$

16. $y = \dfrac{4}{3}\csc(3x - 2) + \dfrac{5}{4}$

17. $y = 8\csc\left(\pi x - \dfrac{\pi}{3}\right) + 2$

18. $y = -\sqrt{2}\csc\left(\dfrac{\pi x}{4} + \pi\right) + \sqrt{2}$

19. $y = 6\csc\left(\dfrac{x}{3} + 3\right) - 2$

20. $y = -2\csc\left(\dfrac{4x}{3} - 3\right) + 6$

For questions 21 to 24, rewrite the function as an equivalent cosecant function in which the coefficient of x is positive.

21. $y = 3\csc(-2x)$

22. $y = -\csc\left(-3x + \dfrac{\pi}{4}\right)$

23. $y = -4\csc\left(-1.5x - \dfrac{2\pi}{3}\right) + 2$

24. $y = \sqrt{2}\csc\left(-\dfrac{x}{3} + \dfrac{\pi}{6}\right)$

25. Sketch at least one cycle of the graph of $y = \csc(-2x + \pi)$.

The Graph of $y = A\cot(Bx - C) + D$

The function $y = A\cot(Bx - C) + D$ has domain $x \neq \dfrac{C}{B} + \dfrac{n\pi}{B}$, where n is an integer, and range $(-\infty, \infty)$. Its graph has undefined amplitude, stretching factor of $|A|$, period $\dfrac{\pi}{|B|}$, horizontal phase shift $\dfrac{C}{B}$, vertical shift D, and vertical asymptotes at $x = \dfrac{C}{B} + \dfrac{n\pi}{B}$, where n is an integer. Note: If $B < 0$, use your knowledge of odd functions to rewrite $y = A\cot(Bx - C) + D$ as the equivalent function $y = -A\cot(|B|\,x + C) + D$.

The graph of $y = \dfrac{1}{3}\cot\left(2x + \dfrac{3\pi}{2}\right) + 1$ shown below has period $\dfrac{\pi}{2}$,

stretching factor $\dfrac{1}{3}$, a horizontal shift of $\dfrac{3\pi}{4}$ to the left, and a vertical shift of

1 unit upward.

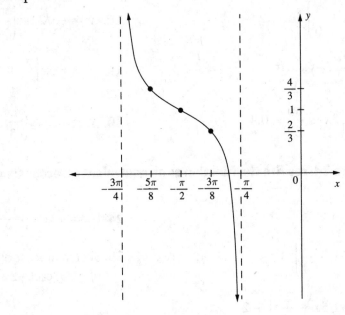

EXERCISE 12-3

For questions 1 to 5, determine the stretching factor and the period of the given function.

1. $y = 3\cot\left(x - \dfrac{\pi}{4}\right) - 2$

2. $y = -5\cot\left(x + \dfrac{\pi}{4}\right) + 1$

3. $y = 4\cot\left(2x + \dfrac{\pi}{6}\right) + \dfrac{1}{2}$

4. $y = \dfrac{1}{2}\cot(2x - \pi) + \sqrt{5}$

5. $y = -0.6\cot\left(3x - \dfrac{\pi}{2}\right) - 0.4$

For questions 6 to 10, describe the horizontal and vertical shifts of the graph of the given function.

6. $y = \dfrac{4}{3}\cot(3x - 2) + \dfrac{5}{4}$

7. $y = 8\cot\left(\pi x - \dfrac{\pi}{3}\right) + 2$

8. $y = -\sqrt{2}\cot\left(\dfrac{\pi x}{4} + \pi\right) + \sqrt{2}$

9. $y = 6\cot\left(\dfrac{x}{3} + 3\right) - 2$

10. $y = -2\cot\left(\dfrac{4x}{3} - 3\right) + 6$

For questions 11 to 20, determine the asymptotes for the graph of the given function.

11. $y = 3\cot\left(x - \dfrac{\pi}{4}\right) - 2$

12. $y = -5\cot\left(x + \dfrac{\pi}{4}\right) + 1$

13. $y = 4\cot\left(2x + \dfrac{\pi}{6}\right) + \dfrac{1}{2}$

14. $y = \dfrac{1}{2}\cot(2x - \pi) + \sqrt{5}$

15. $y = -0.6\cot\left(3x - \dfrac{\pi}{2}\right) - 0.4$

16. $y = \dfrac{4}{3}\cot(3x - 2) + \dfrac{5}{4}$

17. $y = 8\cot\left(\pi x - \dfrac{\pi}{3}\right) + 2$

18. $y = -\sqrt{2}\cot\left(\dfrac{\pi x}{4} + \pi\right) + \sqrt{2}$

19. $y = 6\cot\left(\dfrac{x}{3} + 3\right) - 2$

20. $y = -2\cot\left(\dfrac{4x}{3} - 3\right) + 6$

For questions 21 to 24, rewrite the function as an equivalent cotangent function in which the coefficient of x is positive.

21. $y = 3\cot(-2x)$

22. $y = -\cot\left(-3x + \dfrac{\pi}{4}\right)$

23. $y = -4\cot\left(-1.5x - \dfrac{2\pi}{3}\right) + 2$

24. $y = \sqrt{2}\cot\left(-\dfrac{x}{3} + \dfrac{\pi}{6}\right)$

25. Sketch at least one cycle of the graph

of $y = \cot\left(x + \dfrac{\pi}{6}\right)$.

CHAPTER 13

Inverse Trigonometric Functions

The Inverse Sine, Cosine, and Tangent Functions

A function is **one-to-one** if every element of the range corresponds to exactly one element of the domain of the function. For a function to have an inverse, the function must be one-to-one. The trigonometric functions are periodic on the real numbers. Consequently, none of the trigonometric functions are one-to-one functions on their domains. However, if their domains are suitably restricted, then each has a unique inverse function.

The restricted domains of the sine, cosine, and tangent functions are given in the following table along with the domains and ranges of their inverse functions (represented using the notations $\sin^{-1} x, \cos^{-1} x,$ and $\tan^{-1} x$). Notice that the outputs of the inverse functions are limited to specific range values. These limitations are a necessary result of the one-to-one function requirement for the function and its inverse.

Other notations for the inverse trigonometric functions use an arc- prefix (e.g., arcsin x).

Restricted Domains of the Sine, Cosine, and Tangent Functions		
$y = \sin x$	$y = \cos x$	$y = \tan x$
$-\dfrac{\pi}{2} \leq x \leq \dfrac{\pi}{2}$	$0 \leq x \leq \pi$	$-\dfrac{\pi}{2} < x < \dfrac{\pi}{2}$
Domains and Ranges of the Inverse Sine, Cosine, and Tangent Functions		
$y = \sin^{-1} x$	$y = \cos^{-1} x$	$y = \tan^{-1} x$
$-1 \leq x \leq 1$ (domain)	$-1 \leq x \leq 1$ (domain)	$-\infty < x < \infty$ (domain)
$-\dfrac{\pi}{2} \leq y \leq \dfrac{\pi}{2}$ (range)	$0 \leq y \leq \pi$ (range)	$-\dfrac{\pi}{2} < y < \dfrac{\pi}{2}$ (range)

The graphs of the trigonometric sine, cosine, and tangent functions on their restricted domains and their corresponding inverses are shown below.

You will find it helpful to read $\theta = \sin^{-1} x$ as "θ is the angle whose sine is x."

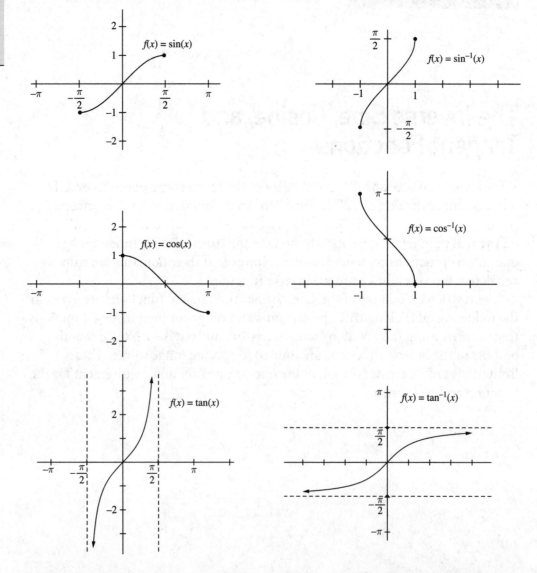

The ranges of the trigonometric functions on their restricted domains are identical to their ranges on the real numbers. Therefore, you can enter any number in a trigonometric function's range as the input for the corresponding inverse trigonometric function. However, the output of the inverse function must lie in a limited range interval that matches up with the restricted domain used to make the original function one-to-one. For example,

$\sin^{-1}\left(\dfrac{1}{2}\right)$ is $\dfrac{\pi}{6}$ because $\dfrac{\pi}{6}$ is the one and only angle in the interval $\left[-\dfrac{\pi}{2}, \dfrac{\pi}{2}\right]$

whose sine is $\dfrac{1}{2}$. As explained in Appendix A, the trigonometric features on calculators use these limited ranges to determine output values for the inverse trigonometric functions.

To summarize:

$\sin^{-1} x$ is the real number (or angle) in the interval $\left[-\dfrac{\pi}{2}, \dfrac{\pi}{2}\right]$ whose sine is x.

$\cos^{-1} x$ is the real number (or angle) in the interval $\left[0, \pi\right]$ whose sine is x.

$\tan^{-1} x$ is the real number (or angle) in the interval $\left(-\dfrac{\pi}{2}, \dfrac{\pi}{2}\right)$ whose sine is x.

> Because the trigonometric functions are periodic, intervals other than those selected for their restricted domains could be chosen; however, mathematicians generally agree on the intervals given in this chapter (as evidenced by their adoption in technology).

> When you evaluate $\sin^{-1} x$, $\cos^{-1} x$, or $\tan^{-1} x$, express the result in radians.

EXAMPLE

▶ Find the exact value of $\sin^{-1}\dfrac{\sqrt{2}}{2}$. (Express the answer in terms of π.)

▶ Because $\sin\dfrac{\pi}{4} = \dfrac{\sqrt{2}}{2}$ and $\dfrac{\pi}{4}$ lies in the interval $\left[-\dfrac{\pi}{2}, \dfrac{\pi}{2}\right]$, then

$\sin^{-1}\dfrac{\sqrt{2}}{2} = \dfrac{\pi}{4}$.

EXAMPLE

▶ Find the exact value of $\cos^{-1}\left(-\dfrac{1}{2}\right)$. (Express the answer in terms of π.)

▶ Because $\cos\dfrac{2\pi}{3} = -\dfrac{1}{2}$ and $\dfrac{2\pi}{3}$ lies in the interval $[0, \pi]$,

then $\cos^{-1}\left(-\dfrac{1}{2}\right) = \dfrac{2\pi}{3}$.

Find the exact value of $\sin\left(\cos^{-1}\dfrac{\sqrt{3}}{2}\right)$.

$$\sin\left(\cos^{-1}\frac{\sqrt{3}}{2}\right) = \sin\frac{\pi}{6} = \frac{1}{2}$$

Find $\sin^{-1}\left(\sin\dfrac{5\pi}{6}\right)$. (Express the answer in terms of π.)

$\sin^{-1}\left(\sin\dfrac{5\pi}{6}\right) = \sin^{-1}\left(\dfrac{1}{2}\right) = \dfrac{\pi}{6}$; because $\sin\dfrac{\pi}{6} = \dfrac{1}{2}$ and $\dfrac{\pi}{6}$ lies in the

interval $\left[-\dfrac{\pi}{2},\dfrac{\pi}{2}\right]$. Caution: $\sin^{-1}\left(\sin\dfrac{5\pi}{6}\right) \neq \dfrac{5\pi}{6}$ because $\dfrac{5\pi}{6}$ does *not* lie in

the interval $\left[-\dfrac{\pi}{2},\dfrac{\pi}{2}\right]$.

Use a calculator in radian mode to find an approximation for
$\tan^{-1}(-0.5906)$. (Round the answer to one decimal place.)

$\text{Tan}^{-1}(-0.5906) \approx -0.5$. Note: You do not have to check whether -0.5

lies in the interval $\left(-\dfrac{\pi}{2},\dfrac{\pi}{2}\right)$ (which, by the way, it does) because calculators

are programmed to output answers consistent with the definitions of the
inverse sine, cosine, and tangent functions.

Simplify $\cos\left(\tan^{-1}\left(\dfrac{\sqrt{x^2-9}}{3}\right)\right)$. (Assume $x > 0$.)

Let $\theta = \tan^{-1}\left(\dfrac{\sqrt{x^2-9}}{3}\right)$. Then, $\tan\theta = \dfrac{\sqrt{x^2-9}}{3}$. Sketch a right

triangle that fits this tangent ratio. Designate $\sqrt{x^2-9}$ as the side
opposite θ and 3 as the side adjacent to θ, making the hypotenuse equal
to $\sqrt{x^2-9+9} = \sqrt{x^2} = x$.

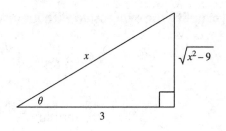

Using the sketch, $\cos\left(\tan^{-1}\left(\dfrac{\sqrt{x^2-9}}{3}\right)\right) = \cos\theta = \dfrac{3}{x}$

EXERCISE 13-1

For questions 1 to 8, find the exact value of each expression. (Express the answer in terms of π.)

1. $\cos^{-1}\dfrac{1}{2}$

2. $\tan^{-1}\sqrt{3}$

3. $\cos^{-1}0$

4. $\sin^{-1}\left(-\dfrac{\sqrt{2}}{2}\right)$

5. $\tan^{-1}\dfrac{1}{\sqrt{3}}$

6. $\tan^{-1}(-\sqrt{3})$

7. $\sin^{-1}\left(-\dfrac{1}{2}\right)$

8. $\cos^{-1}\left(-\dfrac{\sqrt{3}}{2}\right)$

For questions 9 to 15, use a calculator in radian mode to find an approximate real-number value for the given expression. Round the answer to one decimal place.

9. $\sin^{-1}(0.3524)$

10. $\cos^{-1}(0.6705)$

11. $\tan^{-1}(-1.834)$

12. $\sin^{-1}(-0.9834)$

13. $\cos^{-1}(0.8962)$

14. $\tan^{-1}(0.5627)$

15. $\sin^{-1}(0.9991)$

For questions 16 to 20, find the exact value of the given expression. (When applicable, express the answer in terms of π.)

16. $\sin\left(\sin^{-1}\dfrac{\sqrt{3}}{2}\right)$

17. $\tan\left(\cos^{-1}\dfrac{1}{2}\right)$

18. $\sin^{-1}\left(\sin\dfrac{3\pi}{4}\right)$

19. $\cos^{-1}\left(\sin\dfrac{\pi}{6}\right)$

20. $\cos(\tan^{-1}\sqrt{3})$

For questions 21 to 25, simplify the expression without using a calculator.

21. $\sin\left(\cos^{-1}\dfrac{24}{25}\right)$

24. $\sin\left(\cos^{-1}\left(\dfrac{\sqrt{x^2-9}}{x}\right)\right)$

22. $\cos\left(\tan^{-1}\dfrac{1}{\sqrt{3}}\right)$

25. $\tan\left(\sin^{-1}\dfrac{4}{5}\right)$

23. $\sin\left(\tan^{-1}\left(\dfrac{4}{\sqrt{x^2-16}}\right)\right)$

The Inverse Secant, Cosecant, and Cotangent Functions

The inverse secant, cosecant, and cotangent functions are used infrequently, but are included here for completeness. These inverse functions are defined in a manner similar to that used to define the inverses of the sine, cosine, and tangent functions. The following table shows the domains and ranges of the inverse secant, cosecant, and cotangent functions:

Domains and Ranges of the Inverse Secant, Cosecant, and Cotangent Functions		
$y = \sec^{-1} x$	$y = \csc^{-1} x$	$y = \cot^{-1} x$
$\|x\| \geq 1$ (domain)	$\|x\| \geq 1$ (domain)	$-\infty < x < \infty$ (domain)
$0 \leq y \leq \pi, y \neq \dfrac{\pi}{2}$ (range)	$-\dfrac{\pi}{2} \leq y \leq \dfrac{\pi}{2}, y \neq 0$ (range)	$-\dfrac{\pi}{2} < y < 0$ or $0 < y \leq \dfrac{\pi}{2}$ (range)

The graphs of the secant, cosecant, and cotangent functions on their restricted domains and their corresponding inverses are shown below.

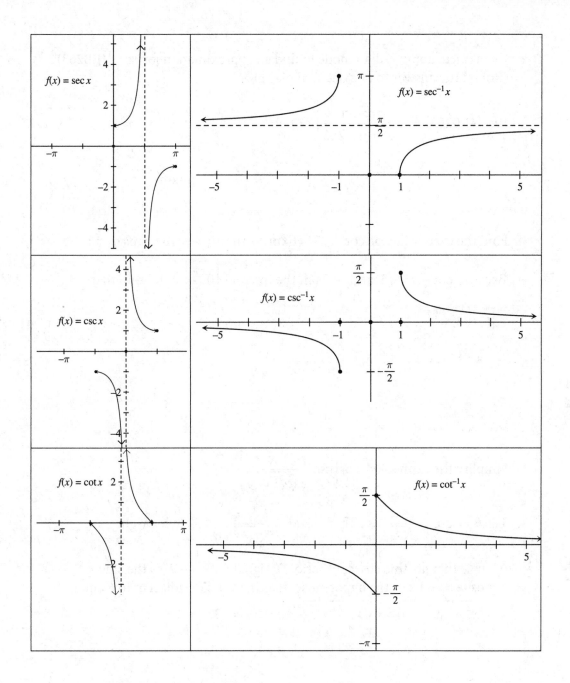

The values of $\sec^{-1} x$, $\csc^{-1} x$, and $\cot^{-1} x$ cannot be found directly on most calculators. However, you can obtain the values by using the following formulas derived from the reciprocal relationships (see Appendix A for instructions):

$$\sec^{-1} x = \cos^{-1} \frac{1}{x} \qquad \csc^{-1} x = \sin^{-1} \frac{1}{x} \qquad \cot^{-1} x = \tan^{-1} \frac{1}{x}$$

EXAMPLE

Use a calculator in radian mode to find an approximation for $\sec^{-1}(1.0263)$. (Round the answer to one decimal place.)

$$\sec^{-1}(1.0263) = \cos^{-1}\left(\frac{1}{1.0263}\right) \approx 0.2$$

EXAMPLE

Find the exact value of $\cot^{-1}\sqrt{3}$. (Express the answer in terms of π.)

Because $\cot\dfrac{\pi}{6} = \sqrt{3}$ and $\dfrac{\pi}{6}$ is in the interval $\left[0, \dfrac{\pi}{2}\right]$, it follows that

$$\cot^{-1}\sqrt{3} = \frac{\pi}{6}.$$

EXAMPLE

Simplify the expression $\cos\left(\csc^{-1}\dfrac{x}{\sqrt{x^2-9}}\right)$.

Let $\theta = \csc^{-1}\dfrac{x}{\sqrt{x^2-9}}$. Then, $\csc\theta = \dfrac{x}{\sqrt{x^2-9}}$. Sketch a right triangle that fits this cosecant ratio. Designate $\sqrt{x^2-9}$ as the side opposite θ and x as the hypotenuse, making the side adjacent to θ equal to $\sqrt{x^2-(x^2-9)} = \sqrt{x^2-x^2+9} = \sqrt{9} = 3$.

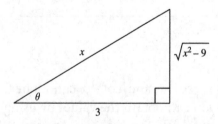

Using the sketch, $\cos\left(\csc^{-1}\dfrac{x}{\sqrt{x^2-9}}\right) = \cos\theta = \dfrac{3}{x}$.

EXERCISE 13-2

For questions 1 to 5, find the exact value of each expression. (Express the answer in terms of π.)

1. $\sec^{-1} \dfrac{2}{\sqrt{3}}$

2. $\csc^{-1} 2$

3. $\cot^{-1} 1$

4. $\sec^{-1} \sqrt{2}$

5. $\csc^{-1} \dfrac{2}{\sqrt{3}}$

For questions 6 to 10, use a calculator in radian mode to find an approximate real-number value for the given expression. Round the answer to one decimal place.

6. $\csc^{-1} (1.1067)$

7. $\sec^{-1} (1.0436)$

8. $\cot^{-1} (1.2356)$

9. $\csc^{-1} (1.3692)$

10. $\sec^{-1} (1.3657)$

For questions 11 to 15, find the exact value of each expression. (When applicable, express the answer in terms of π.)

11. $\sec\left(\sec^{-1} \dfrac{2}{\sqrt{3}}\right)$

12. $\cos(\cot^{-1} \sqrt{3})$

13. $\csc^{-1}\left(\csc \dfrac{\pi}{4}\right)$

14. $\sec^{-1}\left(\csc \dfrac{\pi}{6}\right)$

15. $\tan\left(\sec^{-1} \dfrac{2}{\sqrt{3}}\right)$

For questions 16 to 20, simplify the expression without using a calculator.

16. $\tan\left(\sec^{-1}\left(\dfrac{\sqrt{9 + x^2}}{x}\right)\right)$

17. $\cos\left(\cot^{-1} \dfrac{x}{3}\right)$

18. $\sin\left(\sec^{-1}\left(\dfrac{\sqrt{x^2 + 25}}{5}\right)\right)$

19. $\sec(\cot^{-1} x)$

20. $\csc\left(\sec^{-1}\left(\dfrac{\sqrt{x^2 + 4}}{x}\right)\right)$

CHAPTER 14

Solving Trigonometric Equations

Basic Concepts of Trigonometric Equations

A **trigonometric equation** is an equation in which the variable is the argument of one or more trigonometric functions. In Chapter 7, you learned about trigonometric identities; that is, equations that are true for all real number (or angle) replacements of the variable for which all functions involved are defined. In this chapter, you will learn about **conditional trigonometric equations**; that is, equations that are true for only particular real number (or angle) replacements of the variable.

The key to solving trigonometric equations is the exploitation of the following functional relationships of one-to-one functions:

Let f be a one-to-one function, then $y = f(x) \Leftrightarrow x = f^{-1}(y)$. Equivalently, $f(f^{-1}(y)) = y \Leftrightarrow f^{-1}(f(x)) = x$.

Caution: The inverse functional relationships work only for x in the proper domain. For example, because the $\cos^{-1}$ function is defined by restricting the domain of the cosine function to $[0, \pi]$, $\cos^{-1}\left(\cos\dfrac{5\pi}{3}\right) \neq \dfrac{5\pi}{3}$. This situation is due to the fact that $\dfrac{5\pi}{3}$ is not in the interval $[0, \pi]$. However, $\cos\left(\dfrac{5\pi}{3}\right) = \dfrac{1}{2}$, and $\dfrac{1}{2}$ is a permissible input for the $\cos^{-1}$ function because it lies in the interval $[-1, 1]$, which is the domain of the $\cos^{-1}$ function.

> [The symbol "$\Leftrightarrow$" means "if and only if."]

135

In the restricted domain of the cosine function, the one and only value for which $\cos x = \dfrac{1}{2}$ is $\dfrac{\pi}{3}$. Therefore, because the $\cos^{-1}$ function is a one-to-one function, it recognizes $\cos^{-1}\left(\dfrac{1}{2}\right)$ only as $\cos^{-1}\left(\cos\dfrac{\pi}{3}\right)$. With the argument of the cosine function in its restricted domain, you can apply $f^{-1}(f(x)) = x$ to obtain $\cos^{-1}\left(\cos\dfrac{\pi}{3}\right) = \dfrac{\pi}{3}$. As you are aware, there are infinitely many values for which $\cos x = \dfrac{1}{2}$ but only the ordered pair $\left(\dfrac{1}{2}, \dfrac{\pi}{3}\right)$ is an element of the $\cos^{-1}$ function.

In practice, you simply have to be mindful of the following restrictions when evaluating $\sin^{-1} x$, $\cos^{-1} x$, and $\tan^{-1} x$:

▶ For $\sin^{-1} x$, the output must lie in the interval $\left[-\dfrac{\pi}{2}, \dfrac{\pi}{2}\right]$.

▶ For $\cos^{-1} x$, the output must lie in the interval $[0, \pi]$.

▶ For $\tan^{-1} x$, the output must lie in the interval $\left(-\dfrac{\pi}{2}, \dfrac{\pi}{2}\right)$.

The following table showing the algebraic signs of the trigonometric functions in each quadrant is a valuable tool to help in solving trigonometric equations.

Signs of the Trigonometric Ratios, x ≠ 0 and y ≠ 0				
Quadrant	I	II	III	IV
$\sin\theta$	+	+	−	−
$\cos\theta$	+	−	−	+
$\tan\theta$	+	−	+	−

If $x = 0$ or $y = 0$, then the angles involved are quadrantal angles and the trigonometric functions take on the special values of 0, 1, −1, or are undefined. You will draw on your knowledge of the function values of quadrantal angles to solve equations involving quadrantal angles as special cases (see Chapter 6 for reference).

Here are guidelines for solving trigonometric equations:

1. If the equation contains different trigonometric functions, if possible, use identities and algebraic manipulation (reducing fractions, adding/subtracting fractions, factoring, etc.) to express the different functions in terms of the same function.

2. If the arguments of the trigonometric functions are different but related measures, use identities to write all functions in terms of the same argument.

Making errors such as $\cos^{-1}\left(\cos\dfrac{5\pi}{3}\right)$ $= \dfrac{5\pi}{3}$ or $\cos^{-1}\left(\dfrac{1}{2}\right)$ $= \dfrac{5\pi}{3}$ is a common pitfall for students of trigonometry. Remembering the restricted domains that are used to create the inverse trigonometric functions can help you avoid such mistakes.

Note that none of the inverse trigonometric functions returns an output that lies in quadrant III.

3. After all functions are expressed in terms of the same trigonometric function with the same argument, isolate the function and then use your understanding of inverse functions to solve for the variable. If the equation has a quadratic form, first use factoring or the quadratic formula to solve it for the trigonometric function involved.

Keep in mind that not all trigonometric equations have solutions, but solving those that do requires applying your knowledge of trigonometry from previous chapters in this workbook and using previously acquired algebraic manipulation skills.

EXAMPLE

▶ Solve $\sin\theta = \dfrac{1}{2}$.

In terms of radian measure, if $\sin\theta = \dfrac{1}{2}$ and $0 \le \theta \le 2\pi$, then

$$\sin^{-1}(\sin\theta) = \sin^{-1}\left(\dfrac{1}{2}\right) = \dfrac{\pi}{6} \text{ implies } \theta = \dfrac{\pi}{6} \text{ (quadrant I) or}$$

$\theta = \pi - \dfrac{\pi}{6} = \dfrac{5\pi}{6}$ (quadrant II). Any value of θ that is coterminal with $\dfrac{\pi}{6}$

or $\dfrac{5\pi}{6}$ is also a solution. The general solution of $\sin\theta = \dfrac{1}{2}$ in radians is

$\dfrac{\pi}{6} + n \cdot 2\pi$ or $\dfrac{5\pi}{6} + n \cdot 2\pi$, where n is an integer.

▶ In terms of degree measure, if $\sin\theta = \dfrac{1}{2}$ and $0° \le \theta \le 360°$,

$$\sin^{-1}(\sin\theta) = \sin^{-1}\left(\dfrac{1}{2}\right) = 30° \text{ implies } \theta = 30° \text{ (quadrant I) or}$$

$\theta = 180° - 30° = 150°$ (quadrant II). Any value of θ that is coterminal with

$30°$ or $150°$ is also a solution. The general solution of $\sin\theta = \dfrac{1}{2}$ in degrees is

$30° + n \cdot 360°$ or $150° + n \cdot 360°$, where n is an integer.

EXAMPLE

▶ Solve $\tan\theta = \cot\theta$ for $0 \le \theta \le 2\pi$.

▶ Using the reciprocal identity $\cot\theta = \dfrac{1}{\tan\theta}$, transform the equation to

$\tan\theta = \dfrac{1}{\tan\theta}$ and simplify.

$$\tan\theta = \dfrac{1}{\tan\theta}$$
$$\tan^2\theta = 1$$
$$\tan^2\theta - 1 = 0$$

▶ This result is a quadratic equation in terms of $\tan\theta$. First, solve for $\tan\theta$ and then solve for θ.

$$\tan^2\theta - 1 = 0$$
$$(\tan\theta + 1)(\tan\theta - 1) = 0$$
$$\tan\theta = -1 \text{ or } \tan\theta = 1$$

▶ If $\tan\theta = -1$, then $\tan^{-1}(\tan\theta) = \tan^{-1}(-1) = -\dfrac{\pi}{4}$ implies $\theta = \dfrac{3\pi}{4}$ (quadrant II) or $\theta = \dfrac{7\pi}{4}$ (quadrant IV).

▶ If $\tan\theta = 1$, then $\tan^{-1}(\tan\theta) = \tan^{-1}(1) = \dfrac{\pi}{4}$ implies $\theta = \dfrac{\pi}{4}$ (quadrant I) or $\theta = \dfrac{5\pi}{4}$ (quadrant III).

▶ Hence, the complete solution is $\theta = \dfrac{\pi}{4}, \dfrac{3\pi}{4}, \dfrac{5\pi}{4}$, or $\dfrac{7\pi}{4}$.

EXAMPLE

▶ Solve $\cos 2\theta - \sin\theta = 0$ for $0 \leq \theta \leq \dfrac{\pi}{2}$.

▶ Using the double-angle identity $\cos 2\theta = 1 - 2\sin^2\theta$, transform the equation to $1 - 2\sin^2\theta - \sin\theta = 0$ and simplify.

$$1 - 2\sin^2\theta - \sin\theta = 0$$
$$2\sin^2\theta + \sin\theta - 1 = 0$$

▶ This result is a quadratic equation in terms of $\sin\theta$. First, solve for $\sin\theta$ and then solve for θ.

$$2\sin^2\theta + \sin\theta - 1 = 0$$
$$(2\sin\theta - 1)(\sin\theta + 1) = 0$$
$$\sin\theta = \dfrac{1}{2} \text{ or } \sin\theta = -1$$

▶ If $\sin\theta = \dfrac{1}{2}$, then $\sin^{-1}(\sin\theta) = \sin^{-1}\left(\dfrac{1}{2}\right) = \dfrac{\pi}{6}$ implies $\theta = \dfrac{\pi}{6}$, which lies in the interval $\left[0, \dfrac{\pi}{2}\right]$.

▶ The equation $\sin\theta = -1$ is rejected because it does not have a solution for which θ lies in the interval $\left[0, \dfrac{\pi}{2}\right]$.

▶ Hence, the solution is $\theta = \dfrac{\pi}{6}$.

EXERCISE 14-1

For questions 1 to 6, state whether the equation is a conditional equation (C) or an identity (I).

1. $2\cos\theta + 1 = 0$

2. $\tan\theta = \dfrac{\sin\theta}{\cos\theta}$

3. $\cos^2\theta - \sin^2\theta = 1$

4. $\tan 2\theta = \dfrac{2\tan\theta}{1 - \tan^2\theta}$

5. $\sin\theta\tan\dfrac{\theta}{2} = \dfrac{2 - \sqrt{2}}{2}$

6. $\sec 3\theta = 2$

For questions 7 to 10, solve for θ in the indicated interval without the use of the trigonometric features on technological devices.

7. $2\cos\theta + 1 = 0$, $0° \leq \theta \leq 360°$

8. $\cos^2\theta - \sin^2\theta = \dfrac{\sqrt{3}}{2}$, $0° \leq \theta \leq 180°$

9. $\sin\theta\tan\dfrac{\theta}{2} = \dfrac{2 - \sqrt{2}}{2}$, $0 \leq \theta \leq 2\pi$

10. $\sec 3\theta = 2$, $0 \leq \theta \leq \dfrac{2\pi}{3}$

Solving for Exact Solutions to Trigonometric Equations

Solving for exact solutions to trigonometric equations requires a good working knowledge of the trigonometric values of the special angles:

$$0°, 30°, 45°, 60°, 90°, 180°, 270°, \text{ and } 360°$$

and their radian counterparts:

$$0, \frac{\pi}{6}, \frac{\pi}{4}, \frac{\pi}{3}, \frac{\pi}{2}, \pi, \frac{3\pi}{2}, \text{ and } 2\pi$$

EXAMPLE

Solve the equation $\tan\theta = 1$ for $0° \leq \theta \leq 360°$.

If $\tan\theta = 1$, then $\tan^{-1}(\tan\theta) = \tan^{-1}(1) = 45°$ implies $\theta = 45°$ (quadrant I) or $\theta = 225°$ (quadrant III).

Hence, the solution is $\theta = 45°$ or $225°$.

EXAMPLE

Solve the equation $\tan 4\theta = 1$ for $0 \leq \theta \leq \dfrac{\pi}{2}$.

If $\tan 4\theta = 1$, then $\tan^{-1}(\tan 4\theta) = \tan^{-1}1 = \dfrac{\pi}{4}$ implies $4\theta = \dfrac{\pi}{4}$

(quadrant I) or $4\theta = \dfrac{5\pi}{4}$ (quadrant III) in which case $\theta = \dfrac{\pi}{16}$ or $\theta = \dfrac{5\pi}{16}$,

both of which lie in the interval $0 \leq \theta \leq \dfrac{\pi}{2}$. Hence, the solution is

$\theta = \dfrac{\pi}{16}$ or $\theta = \dfrac{5\pi}{16}$.

EXERCISE 14-2

For questions 1 to 10, solve the equations for the exact solution value(s) for $0 \leq \theta \leq \dfrac{\pi}{2}$ without the use of the trigonometric features on technological devices.

1. $\cos 3\theta = 1$

2. $\tan 4\theta = -\sqrt{3}$

3. $\sin 3\theta = 0$

4. $\sin 2\theta = \cos 2\theta$

5. $\cot 4\theta = -1$

6. $\cos 5\theta = 1$

7. $\sin 5\theta = \dfrac{\sqrt{2}}{2}$

8. $\sin \dfrac{\theta}{2} = \tan \dfrac{\theta}{2}$

9. $\sin 2\theta - \cos\theta = 0$

10. $\cos 2\theta + \cos\theta = 0$

For questions 11 to 20, solve the equations for the exact solution value(s) for $0° \leq \theta \leq 90°$ without the use of the trigonometric features on technological devices.

11. $\sin 4\theta = \dfrac{1}{2}$

12. $\sec 3\theta = \dfrac{2}{\sqrt{3}}$

13. $\cot 5\theta = 1$

14. $\cos 4\theta = \dfrac{1}{2}$

15. $2\cos\theta - 1 = 0$

16. $5\cos\theta - 2\sqrt{3} = \sqrt{3} - \cos\theta$

17. $16\sin^2(2\theta) = 4$

18. $2\cos^2\theta - 2\sin^2\theta = \sqrt{3}$

19. $4\cos\theta\sin\theta = \sqrt{2}$

20. $\tan^2(2\theta) - 2\sqrt{3}\tan 2\theta = -3$

Solving for Approximate Solutions to Trigonometric Equations

Use the inverse trigonometric functions on a calculator to get approximate solutions for trigonometric equations involving specific numerical values. The inverse trigonometric functions have restricted outputs. The restrictions are consistent with inverse function calculations with calculators (and similar technological devices).

When the calculator returns a value, that output determines a reference angle that you can use to obtain the solution(s) (see Chapter 6 for reference). Generally, there will be two angles between $0°$ and $360°$ (0 and 2π) that are possible solutions to an equation (see Appendix A for a discussion of trigonometric calculator use).

> When using trigonometric calculator keys, entering an input value for a trigonometric function that is not in the domain of the trigonometric function, will result in an error message.

EXAMPLE

▶ Solve the equation $\sin\theta = 0.7346$ for $0° \leq \theta \leq 360°$.

▶ Given $\sin\theta = 0.7346$, then θ is an angle whose sine is 0.7346. Because sine is positive in quadrants I and II, there are two values for θ in the given interval that satisfy the equation. Using the calculator, you get $\theta = \sin^{-1}(0.7346) \approx 47.3°$ as the quadrant I value for θ. Given that $47.3°$ is the reference angle for θ in quadrant II, the other approximate value for θ is $180° - 47.3° = 132.7°$. Thus, $\theta \approx 47.3°$ or $132.7°$.

EXAMPLE

▶ Solve the equation $\cos\theta = -0.4356$ for $0° \leq \theta \leq 360°$.

▶ Given $\cos\theta = -0.4356$, then θ is an angle whose cosine is -0.4356. Because cosine is negative in quadrants II and III, there are two values for θ in the given interval that satisfy the equation. Using the calculator, you get $\cos^{-1}(-0.4356) \approx 115.8°$ as the quadrant II angle that has the given cosine value, with a reference angle of $180° - 115.8° = 64.2°$. Given that $64.2°$ is the reference angle for θ in quadrant III, the other approximate value for θ is $180° + 64.2° = 244.2°$. Thus, $\theta \approx 115.8°$ or $244.2°$.

EXAMPLE

▶ Solve the equation $\sin\theta = 1.3682$.

▶ $\sin^{-1}(1.3682)$ gives an error message since 1.3682 is not in the domain of the inverse sine function. There is no solution.

EXERCISE 14-3

Solve the equations for $0° \leq \theta \leq 360°$. Use of the trigonometric features on calculators is permitted. (Round answers to one decimal place.)

1. $\sin\theta = 0.2419$

2. $\cos\theta = 0.5150$

3. $\tan\theta = -2.6051$

4. $\cos\theta = -0.9085$

5. $\sin\theta = -0.2246$

6. $\tan\theta = 0.3142$

7. $\sin\theta = -0.4567$

8. $\cos\theta = 0.5988$

9. $\tan\theta = 0.3241$

10. $\cos\theta = 0.9321$

CHAPTER 15

Trigonometric Form of a Complex Number

Definition of the Trigonometric Form of a Complex Number

A **complex number** is an expression $a + bi$, where a and b are real numbers and $i = \sqrt{-1}$. The use of trigonometry in the study of complex numbers often simplifies the process of computing products, quotients, and powers of complex numbers. A complex number written $z = a + bi$ is in **rectangular form** (also called **standard form**) since it can be graphed on the rectangular **complex plane** (see Appendix C for a brief discussion of the complex plane). You can obtain the **trigonometric form** of z by constructing the associated right triangle as shown in the figure below.

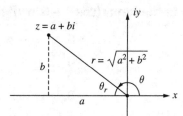

From the figure, $a = r\cos\theta$, $b = r\sin\theta$, $r = \sqrt{a^2 + b^2}$, and $\tan\theta = \dfrac{b}{a}$.

Therefore, the trigonometric form of z is $z = r\cos\theta + r\sin\theta i = r(\cos\theta + i\sin\theta)$.
The positive number r is the **modulus** of z and θ is the **argument** of z.

143

The expression $r(\cos\theta + i\sin\theta)$ is the most commonly used form. Usually $0 \le \theta < 2\pi$, but at times it might be more convenient to choose some other coterminal angle. For instance, the argument θ of a complex number in quadrant IV or on the negative y-axis might be expressed as a negative angle in the interval $\left[-\dfrac{\pi}{2}, 0\right]$.

A shorthand notation for $r(\cos\theta + i\sin\theta)$ is $r\,\text{cis}\,\theta$.

EXAMPLE

> Write $z = 1 + i$ in the form $r(\cos\theta + i\sin\theta)$ with $0 \le \theta < 2\pi$.

> Graphically, z is located in quadrant I of the complex plane with coordinates (1,1).

> $r = \sqrt{x^2 + y^2} = \sqrt{1^2 + 1^2} = \sqrt{2}$, $\tan\theta = \dfrac{y}{x} = \dfrac{1}{1} = 1$, and because z is in quadrant I, $\theta = \tan^{-1}1 = 45°$.

> Thus, $z = \sqrt{2}(\cos 45° + i\sin 45°)$.

EXAMPLE

> Write $z = \sqrt{3} - 1i$ in the form $r(\cos\theta + i\sin\theta)$ with $0 \le \theta < 2\pi$.

> Graphically, z is located in quadrant IV of the complex plane with coordinates $(\sqrt{3}, -1)$.

> $r = \sqrt{x^2 + y^2} = \sqrt{(\sqrt{3})^2 + (-1)^2} = 2$, $\tan\theta = \dfrac{y}{x} = -\dfrac{1}{\sqrt{3}}$, and because z is in quadrant IV, $\theta = \tan^{-1}\left(-\dfrac{1}{\sqrt{3}}\right) = -30°$.

> Thus, $z = 2(\cos(-30°) + i\sin(-30°))$ or $2(\cos 330° + i\sin 330°)$.

EXERCISE 15-1

Write the complex number in the form $r(\cos\theta + i\sin\theta)$ with $0 \le \theta < 2\pi$.

1. $z = 1 + \sqrt{3}i$
2. $z = -1 + i$
3. $z = -\sqrt{3} - 1i$
4. $z = 1$
5. $z = 2 + 2\sqrt{3}i$
6. $z = -1$
7. $z = i$
8. $z = -i$
9. $z = 2 - 2i$
10. $z = \sqrt{3} - i$

The Product and Quotient of Trigonometric Forms of Complex Numbers

If $z_1 = r_1(\cos\theta + i\sin\theta)$ and $z_2 = r_2(\cos\varphi + i\sin\varphi)$, then the product, $z_1 z_2$, is obtained as follows:

$$z_1 z_z = r_1(\cos\theta + i\sin\theta)r_2(\cos\varphi + i\sin\varphi)$$
$$= r_1 r_2(\cos\theta + i\sin\theta)(\cos\varphi + i\sin\varphi)$$
$$= r_1 r_2([\cos\theta\cos\varphi + i^2\sin\theta\sin\varphi] + i[\sin\theta\cos\varphi + \cos\theta\sin\varphi])$$
$$= r_1 r_2([\cos\theta\cos\varphi - \sin\theta\sin\varphi] + i[\sin\theta\cos\varphi + \cos\theta\sin\varphi])$$
$$= r_1 r_2[\cos(\theta + \varphi) + i\sin(\theta + \varphi)]$$

Note the use of the sum formulas for sine and cosine from Chapter 7 in the last line.

In a similar manner, the trigonometric form of the quotient, $\dfrac{z_1}{z_2}$, can be derived. Thus, you have the following two formulas:

$$z_1 z_2 = r_1 r_2[\cos(\theta + \varphi) + i\sin(\theta + \varphi)]$$
$$\frac{z_1}{z_2} = \frac{r_1}{r_2}[\cos(\theta - \varphi) + i\sin(\theta - \varphi)], z_2 \neq 0$$

Note that, according to these formulas, to multiply two complex numbers, you multiply their moduli and add their arguments; to divide two complex numbers, you divide their moduli and subtract their arguments.

The word *moduli* is the plural form of *modulus*.

EXAMPLE

Compute the product of $z_1 = 7(\cos 120° + i\sin 120°)$

and $z_2 = 2(\cos 300° + i\sin 300°)$. Express the result in $a + bi$ form.

$$z_1 z_2 = 7 \cdot 2(\cos(120° + 300°) + i\sin(120° + 300°))$$
$$= 14(\cos 420° + i\sin 420°) = 14(\cos 60° + i\sin 60°)$$

$$= 14 \cdot \cos 60° + 14 \cdot i\sin 60° = 14 \cdot \frac{1}{2} + 14 \cdot i\frac{\sqrt{3}}{2} = 7 + 7\sqrt{3}\,i$$

EXAMPLE

Compute the quotient of $z_1 = 7(\cos 120° + i \sin 120°)$

and $z_2 = 2(\cos 300° + i \sin 300°)$. Express the result in $a + bi$ form.

$$\frac{z_1}{z_2} = \frac{7}{2}(\cos(120° - 300°) + i \sin(120° - 300°))$$

$$= \frac{7}{2}(\cos(-180°) + i \sin(-180°))$$

$$= \frac{7}{2} \cdot \cos(-180°) + \frac{7}{2} \cdot i \sin(-180°) = \frac{7}{2} \cdot (-1) + \frac{7}{2} \cdot i(0) = -\frac{7}{2}$$

EXAMPLE

Compute the product of $z_1 = 3(\cos 70° + i \sin 70°)$

and $z_2 = 4(\cos 38° + i \sin 38°)$. Express the result in $a + bi$ form. (Round coefficients to one decimal place, as needed.)

$$z_1 z_2 = 3 \cdot 4(\cos(70° + 38°) + i \sin(70° + 38°)) = 12(\cos 108° + i \sin 108°)$$

$$= 12 \cdot \cos 108° + 12 \cdot i \sin 108° \approx -3.7 + 11.4i.$$

Complex numbers have many applications in the real world. For example, in electromagnetism, Ohm's Law says that in an electric circuit, the current I in amperes, the voltage V in volts, and the resistance R in ohms are related according to the formulas:

$$I = \frac{V}{R}, V = IR, \text{ and } R = \frac{V}{I}.$$

EXAMPLE

In an electric circuit, find the voltage V if the current

$I = 5(\cos 30° + i \sin 30°)$ and the resistance $R = 2(\cos 15° + i \sin 15°)$. Express the answer in rectangular form.

$$V = IR = [5(\cos 30° + i \sin 30°)][2(\cos 15° + i \sin 15°)]$$

$$= 10(\cos 45° + i \sin 45°) = 5\sqrt{2} + 5\sqrt{2}i$$

EXAMPLE

> In an electric circuit, find the current I if the voltage
> $V = 80(\cos 25° + i\sin 25°)$ and the resistance $R = 40(\cos 35° + i\sin 35°)$.
> Leave the answer in trigonometric form.
>
> $$I = \frac{V}{R} = \frac{80(\cos 25° + i\sin 25°)}{40(\cos 35° + i\sin 35°)} = 2(\cos(-10°) + i\sin(-10°))$$
> $$= 2(\cos(350°) + i\sin(350°)) = 2\text{cis}350°$$

EXERCISE 15-2

For questions 1 to 10, use the trigonometric form to compute the product z_1z_2. Express the result in $a + bi$ form using exact coefficients, when possible; otherwise, round coefficients to one decimal place, as needed.

1. $z_1 = \sqrt{3} + i$ and $z_2 = 1 + \sqrt{3}i$

2. $z_1 = 10(\cos 60° + i\sin 60°)$ and

  $z_2 = 4(\cos 30° + i\sin 30°)$

3. $z_1 = 5\sqrt{2}\ \text{cis}210°$ and $z_2 = 2\sqrt{2}\ \text{cis}30°$

4. $z_1 = 9(\cos\dfrac{\pi}{15} + i\sin\dfrac{\pi}{15})$ and

  $z_2 = 1.8(\cos\dfrac{2\pi}{3} + i\sin\dfrac{2\pi}{3})$

5. $z_1 = 6\text{cis}\ 82°$ and $z_2 = 1.5\text{cis}\ 27°$

6. $z_1 = \sqrt{3} - i$ and $z_2 = 1 + \sqrt{3}i$

7. $z_1 = 1 + i$ and $z_2 = 2 + 2i$

8. $z_1 = \sqrt{3} + i$ and $z_2 = \sqrt{3} + i$

9. $z_1 = i$ and $z_2 = -i$

10. $z_1 = 1 + i$ and $z_2 = -1 + i$

For questions 11 to 15, use the trigonometric form to compute the quotient $\dfrac{z_1}{z_2}$. Express the result in $a + bi$ form using exact coefficients, when possible; otherwise, round coefficients to one decimal place, as needed.

11. $z_1 = 10(\cos 60° + i\sin 60°)$ and

  $z_2 = 4(\cos 30° + i\sin 30°)$

12. $z_1 = 5\sqrt{2}\ \text{cis}210°$ and $z_2 = 2\sqrt{2}\ \text{cis}30°$

13. $z_1 = 9(\cos\dfrac{\pi}{15} + i\sin\dfrac{\pi}{15})$ and

  $z_2 = 1.8(\cos\dfrac{2\pi}{3} + i\sin\dfrac{2\pi}{3})$

14. $z_1 = 6\text{cis}\ 82°$ and $z_2 = 1.5\text{cis}\ 27°$

15. $z_1 = 4(\cos 30° + i\sin 30°)$ and

  $z_2 = 8(\cos 120° + i\sin 120°)$

For questions 16 and 17, solve as indicated.

16. In an electric circuit, find the voltage V if the current $I = 4(\cos 35° + i\sin 35°)$ and the resistance $R = 3(\cos 25° + i\sin 25°)$. Express the answer in rectangular form.

17. In an electric circuit, find the current I if the voltage $V = 100(\cos 45° + i\sin 45°)$ and the resistance $R = 25(\cos 60° + i\sin 60°)$. Leave the answer in trigonometric form.

De Moivre's Theorem

De Moivre's theorem states the following:

If $z = r(\cos\theta + i\sin\theta)$ and n is a positive integer,
then $z^n = r^n[\cos(n\theta) + i\sin(n\theta)]$.

A note of practical information: When n is large, since $\theta \pm 2\pi$ is coterminal with θ, you can use any coterminal angle for the evaluation.

EXAMPLE

Compute $(-2 + 2i)^6$. Express the result in $a + bi$ form.

$-2 + 2i$ is in quadrant II of the complex plane with coordinates $(-2,2)$.

$r = \sqrt{x^2 + y^2} = \sqrt{(-2)^2 + 2^2} = 2\sqrt{2}$ and $\tan\theta = \dfrac{-2}{2} = -1$.

Because $-2 + 2i$ is in quadrant II, $\tan^{-1}(-1) = -45°$ implies $\theta = 135°$.

Thus, $-2 + 2i = 2\sqrt{2}(\cos 135° + i\sin 135°)$. Then,

$(-2 + 2i)^6 = (2\sqrt{2})^6(\cos(6 \cdot 135°) + i\sin(6 \cdot 135°))$

$= (2\sqrt{2})^6(\cos 810° + i\sin 810°)$

$= 512(\cos 90° + i\sin 90°) = 512i$

Given the following definitions, De Moivre's theorem holds for all integers n:
For all $z \neq 0 + 0i$,

▶ $z^0 = 1 + 0i = 1$

▶ $z^{-n} = \dfrac{1}{z^n}$, n a positive integer

EXAMPLE

▶ Compute $(\sqrt{3} - i)^{-3}$. Express the result in $a + bi$ form.

▶ First, substitute the trigonometric form of $\sqrt{3} - i$ into $(\sqrt{3} - i)^{-3}$ and then apply De Moivre's theorem.

$\sqrt{3} - i$ is in quadrant IV of the complex plane with coordinates $(\sqrt{3}, -1)$.

$r = \sqrt{(\sqrt{3})^2 + (-1)^2} = \sqrt{4} = 2$ and $\tan\theta = -\dfrac{1}{\sqrt{3}}$. Because $\sqrt{3} - i$ is in

quadrant IV, $\theta = \tan^{-1}\left(-\dfrac{1}{\sqrt{3}}\right) = -30°$. Then you have

$$(\sqrt{3} - i)^{-3} = \left[2(\cos(-30°) + i\sin(-30°))\right]^{-3}$$
$$= 2^{-3}[\cos(90°) + i\sin(90°)]$$
$$= \frac{1}{8}i$$

EXERCISE 15-3

Use De Moivre's theorem to compute. Express the result in $a + bi$ form using exact coefficients, when possible; otherwise, round coefficients to one decimal place, as needed.

1. $(-1 + i)^6$

2. $[4(\cos 20° + i\sin 20°)]^3$

3. $\left(\dfrac{1}{2} + \dfrac{\sqrt{3}}{2}\right)^3$

4. $[3(\cos 42° + i\sin 42°)]^5$

5. $[16(\cos 315° + i\sin 315)]^{-2}$

6. $(3 + 3i)^6$

7. $(4\cos 300° + 4i\sin 300°)^3$

8. $\left(\dfrac{\sqrt{2}}{2}\,\text{cis}\,135°\right)^8$

9. $\left(\sqrt{3} - i\right)^3$

10. $(4 + 3i)^5$

11. $(1 + i)^{-4}$

12. $\left(\sqrt{3} + \sqrt{3i}\right)^6$

13. $(-3 - 3i)^{-3}$

14. i^{-4}

15. $[5(\cos 15° + i\sin 15°)]^{-4}$

Roots of Complex Numbers

An **nth root of a complex number** z is defined as follows: If z is a complex number, then w is an nth root of z if and only if $w^n = z$. You can apply De Moivre's theorem to find roots of complex numbers using the following theorem:

If n is a positive integer and $z = r(\cos\theta + i\sin\theta)$, $z \neq 0$, then

$$r^{\frac{1}{n}}\left(\cos\left(\frac{\theta + k \cdot 360°}{n}\right) + i\sin\left(\frac{\theta + k \cdot 360°}{n}\right)\right), \text{ where } k \text{ is an integer, specifies}$$

all nth roots of z.

EXAMPLE

Find the three cube roots of $\sqrt{2} + \sqrt{2}i$.

First, express the complex number in polar form.

$r = \sqrt{(\sqrt{2})^2 + (\sqrt{2})^2} = \sqrt{2+2} = 2$, and because $\sqrt{2} + \sqrt{2}i$ is in quadrant

I, $\theta = \tan^{-1}\dfrac{\sqrt{2}}{\sqrt{2}} = \tan^{-1}1 = 45°$. Thus, $\sqrt{2} + \sqrt{2}i = 2(\cos 45° + i\sin 45°)$.

Then the cube roots are given by the formula

$$2^{\frac{1}{3}}\left(\cos\left(\frac{45° + k \cdot 360°}{3}\right) + i\sin\left(\frac{45° + k \cdot 360°}{3}\right)\right), \ k = 0, \pm 1, \pm 2, \pm 3, \dots.$$

Letting $k = 0, 1$, and 2 in succession yields the three roots in polar form.

$2^{\frac{1}{3}}(\cos(15°) + i\sin(15°)), 2^{\frac{1}{3}}(\cos(135°) + i\sin(135°))$, and

$2^{\frac{1}{3}}(\cos(255°) + i\sin(255°))$.

Note: Substituting any other integer for k will yield one of these three values.

In general, for convenience, you can use $k = 0, 1, \dots, n - 1$ to determine the n distinct *nth* roots of a complex number.

EXAMPLE

Solve for the solution set to $z^3 - i = 0$.

Rewrite the equation as $z^3 = i$. Thus, you are to find all cube roots of i.

Express $i = 0 + 1i$ in trigonometric form: $r = 1$ and $\theta = 90°$. Hence, $z^3 = \cos 90° + i\sin 90°$. Using De Moivre's theorem, the solution set to the equation $z^3 = i$ are the three cube roots of $i = \cos 90° + i\sin 90°$, which are obtained as follows:

$$w_k = r^{\frac{1}{3}}\left(\cos\left(\frac{90° + k360°}{3}\right) + i\sin\left(\frac{90° + k360°}{3}\right)\right).$$

Using $k = 0, 1$, and 2 in succession, the three distinct

cube roots of i are $w_0 = \cos 30° + i\sin 30° = \dfrac{\sqrt{3}}{2} + \dfrac{1}{2}i$,

$w_1 = \cos 150° + i\sin 150° = -\dfrac{\sqrt{3}}{2} + \dfrac{1}{2}i$, and

$w_2 = \cos 270° + i\sin 270° = -i$.

As shown below, plotting the three roots on the complex plane shows that they are equally spaced around the origin.

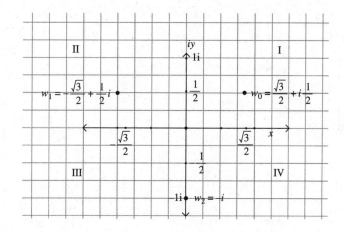

EXERCISE15-4

For questions 1 to 3, find the roots and express them in polar form.

1. The two square roots of $\dfrac{\sqrt{3}}{2} - \dfrac{1}{2}i$

2. The three cube roots of $8(\cos 60° + i \sin 60°)$

3. The four fourth roots of
$81(\cos 30° + i \sin 30°)$

For questions 4 and 5, solve the equation for all complex numbers z and then graph the solution set.

4. $z^3 + 1 = 0$ (Hint: z is a cube root of $-1 = -1 + 0i$)

5. $z^3 - 1 = 0$

CHAPTER 16

Polar Coordinates

Basic Concepts of Polar Coordinates

Polar coordinates are based on a directed distance and a counterclockwise angle relative to a fixed point. As illustrated in the figure below, a fixed ray, called the **polar axis**, emanating from a fixed point, **O**, called the **origin**, is the basis for the coordinate system. A point in the plane is then located by polar coordinates (r,θ).

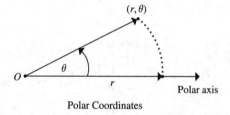

Polar Coordinates

One aspect to this system is that, unlike rectangular coordinates in which the coordinates are unique, a point in the polar system has an unlimited number of representations in polar coordinates. For instance, $(r,\theta) \sim (r,\theta + 2n\pi)$, $n = 1,2,3,...$ [The symbol $\sim$ is used here to mean that representations designate the same point.] Also, $(r,\theta) \sim (-r,\pi + \theta) \sim (-r,-\theta + \pi)$, as illustrated below.

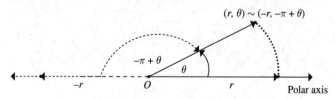

Non-unique representation of polar coordinates

EXAMPLE

Plot the polar coordinates $\left(3, \dfrac{\pi}{4}\right)$, $(2, 30°)$, $(4, 75°)$, $(-3, \pi)$, and $(3, -60°)$.

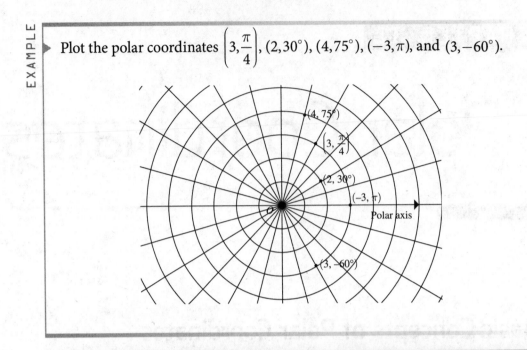

EXERCISE 16-1

Follow the instructions for each of the questions below.

1. Plot the polar coordinates $\left(2, \dfrac{\pi}{3}\right)$, $(3, 90°)$, $(-2, 90°)$, $(-2, 120°)$, and $\left(4, \dfrac{3\pi}{2}\right)$ on the polar grid below.

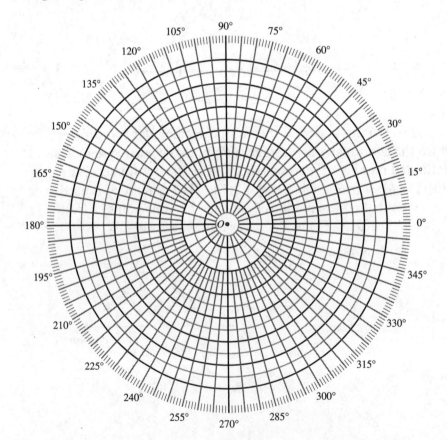

2. Plot the polar coordinates $(2, 405°)$, $(3, -660°)$, and $(-4, -570°)$ on the polar grid below.

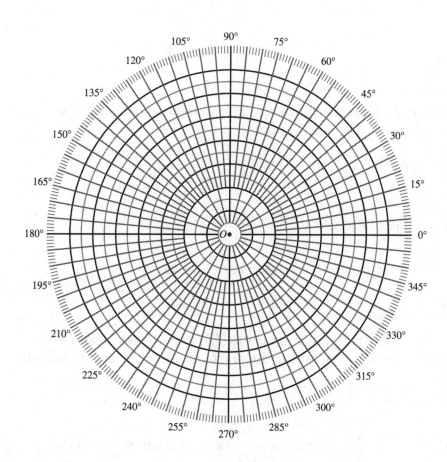

Converting Between Coordinate Systems

Given a point with polar coordinates (r, θ) and rectangular coordinates (x, y), the following conversion formulas are used to convert between the polar and rectangular coordinate systems:

▶ To convert from polar coordinates to rectangular coordinates, use the formulas:

$x = r \cos \theta$ and $y = r \sin \theta$

▶ To convert from rectangular coordinates to polar coordinates, use the formulas:

$r = \sqrt{x^2 + y^2}$ and $\tan \theta = \dfrac{y}{x}, x \neq 0$

Here is an illustration that verifies the formulas.

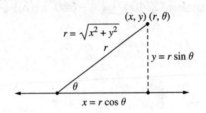

EXAMPLE ▶ Convert the polar coordinates $(6, 240°)$ to rectangular coordinates. Give exact answers. Give exact answers.

$$x = r\cos\theta = 6\cos 240° = 6\left(-\frac{1}{2}\right) = -3$$

$$y = r\sin\theta = 6\sin 240° = 6\left(-\frac{\sqrt{3}}{2}\right) = -3\sqrt{3}$$

▶ Thus, the rectangular coordinates are $(-3, -3\sqrt{3})$.

EXAMPLE ▶ Convert the rectangular coordinates $(3, 6)$ to polar coordinates where $r > 0$ and $0 \le \theta < 360°$. (Round coordinates to one decimal place, as needed.)

$$r = \sqrt{x^2 + y^2} = \sqrt{3^2 + 6^2} = \sqrt{45} \approx 6.7$$

$$\tan\theta = \frac{y}{x} = \frac{6}{3} = 2 \text{, and because } (3,6) \text{ is in quadrant I, } \theta = \tan^{-1} 2 \approx 63.4°$$

▶ Thus, the polar coordinates are $(6.7, 63.4°)$.

EXAMPLE ▶ Convert the rectangular equation $y = 4x$ to an equivalent polar equation.

▶ Substitute the polar forms for the rectangular forms.

$y = 4x$

$r\sin\theta = 4r\cos\theta$

▶ Divide each side by $r\cos\theta$ and simplify.

$r\sin\theta = 4r\cos\theta$

$\dfrac{r\sin\theta}{r\cos\theta} = \dfrac{4r\cos\theta}{r\cos\theta}$

$\dfrac{\sin\theta}{\cos\theta} = 4$

$\tan\theta = 4$

▶ Thus, the equivalent polar equation is $\tan\theta = 4$.

EXAMPLE

▶ Convert the polar equation $r = 6 \csc \theta$ to an equivalent rectangular equation.

▶ Use $\csc \theta = \dfrac{1}{\sin \theta}$ to transform $r = 6 \csc \theta$ to $r = \dfrac{6}{\sin \theta}$.

▶ Multiply each side by $\sin \theta$ to obtain $r \sin \theta = 6$.
Replace $r \sin \theta$ by y to get $y = 6$.
Thus, the equivalent rectangular equation is $y = 6$.

EXERCISE 16-2

For questions 1 to 5, convert from polar coordinates to rectangular coordinates. Give exact answers if possible; otherwise, round to one decimal place, as needed.

1. $\left(4, \dfrac{\pi}{4} \right)$

2. $(6, 750°)$

3. $(5, -450°)$

4. $(-6, -600°)$

5. $\left(3, \dfrac{2\pi}{3} \right)$

For questions 6 to 10, convert from rectangular coordinates to polar coordinates.

6. $(3,0),\ r > 0,\ 0 \le \theta < 2\pi$

7. $(4,4),\ r > 0,\ 0 \le \theta < 2\pi$

8. $(0, -5),\ r > 0,\ 0 \le \theta < 360°$

9. $\left(\dfrac{1}{2}, \dfrac{\sqrt{3}}{2} \right),\ r > 0,\ 0 \le \theta < 360°$

10. $(-\sqrt{3}, 1),\ r > 0,\ 0 \le \theta < 360°$

For questions 11 to 15, convert the rectangular equation to an equivalent polar equation.

11. $x = 5$

12. $y = 4x^2$

13. $x^2 + y^2 = 25$

14. $6x^2 + y^2 = 6$

15. $x^2 + y^2 - 3y = 0$

For questions 16 to 20, convert the polar equation to an equivalent rectangular equation.

16. $r = 8$

17. $r = 2\sec \theta$

18. $r = \dfrac{4}{1 - \sin \theta}$

19. $r = 3\cos \theta$

20. $r = 1 + 2\sin \theta$

Graphing Equations in Polar Form

Graphs of polar equations are plotted in the xy-plane even though the equations themselves are expressed in terms of r and θ. Considering that modern-day calculators, such as the TI-84 Plus, can produce these graphs easily when set to polar mode, this book is written with the assumption that you do not have to sketch them by hand.

Appendix A provides detailed instructions for graphing equations in polar form using the TI-84 Plus calculator. For the curves shown in this section, set your calculator to both radian and polar mode. For a particular equation, it is important that you input window settings that will optimize viewing of the graph. For your convenience, suggested values are given for the equations in this section. For most graphs, you should zoom to a square window to obtain the best results (see Appendix A for instructions).

The graph of $r = a$ is a circle centered at $(0,0)$ with radius a. The graph of $r = 2a \sin \theta$ is a circle with center $(0,a)$ and radius $|a|$. The graph of $r = 2b \cos \theta$ is a circle with center $(b,0)$ and radius $|b|$.

EXAMPLE

Graph $r = 4\sin\theta$.

Window settings:

$\theta\min = 0, \theta\max = 2\pi, \theta\text{step} = \dfrac{\pi}{24}$;

$X\min = -6, X\max = 6, X\text{scl} = 1$;

$Y\min = -4, Y\max = 4, Y\text{scl} = 1$

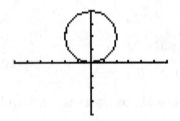

EXAMPLE

Graph $r = 4\cos\theta$.

Window settings:

$\theta\min = 0, \theta\max = 2\pi, \theta\text{step} = \dfrac{\pi}{24}$;

$X\min = -6, X\max = 6, X\text{scl} = 1$;

$Y\min = -4, Y\max = 4, Y\text{scl} = 1$

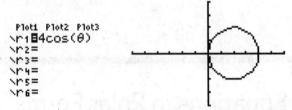

The graphs of $r = a \sin n\theta$ and $r = a \cos n\theta$, where $n > 1$ is an integer, are **rose curves**. If n is odd, there are n petals on the rose; and if n is even, there are $2n$ petals.

EXAMPLE

▶ Graph $r = 4\sin 3\theta$

▶ Window settings:

$\theta\min = 0, \theta\max = 2\pi, \theta\text{step} = \dfrac{\pi}{24};$

$X\min = -6, X\max = 6, X\text{scl} = 1;$

$Y\min = -4, Y\max = 4, Y\text{scl} = 1$

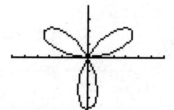

EXAMPLE

▶ Graph $r = 3\cos 2\theta$.

▶ Window settings:

$\theta\min = 0, \theta\max = 2\pi, \theta\text{step} = \dfrac{\pi}{24};$

$X\min = -6, X\max = 6, X\text{scl} = 1;$

$Y\min = -4, Y\max = 4, Y\text{scl} = 1$

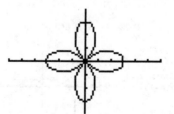

The graphs of $r = a \pm b\sin\theta$ and $r = a \pm b\cos\theta$, where a and b are both positive, are **limaçon curves**.

Limaçon is pronounced LEE-ma-sohn.

EXAMPLE

▶ Graph $r = 2 + 3\cos\theta$.

▶ Window settings:

$\theta\min = 0, \theta\max = 2\pi, \theta\text{step} = \dfrac{\pi}{24}$;

$X\min = -6, X\max = 6, X\text{scl} = 1$;

$Y\min = -4, Y\max = 4, Y\text{scl} = 1$

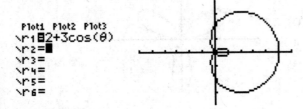

▶ Notice that this limaçon has an inner loop.

EXAMPLE

▶ Graph $r = 3 - 3\sin\theta$.

▶ Window settings:

$\theta\min = 0, \theta\max = 2\pi, \theta\text{step} = \dfrac{\pi}{24}$;

$X\min = -6, X\max = 6, X\text{scl} = 1$;

$Y\min = -8, Y\max = 4, Y\text{scl} = 1$

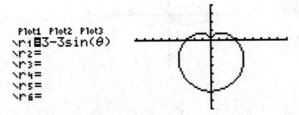

▶ This heart-shaped limaçon is a **cardioid.**

The graphs shown in the examples are only a few of the possible graphs of polar equations. Of course, not all graphs will result in familiar shapes. Nevertheless, the graphs are often interesting and striking.

EXERCISE 16-3

Use a graphing utility to graph the polar equations.

For questions 1 to 7, use the following window settings:

$$\theta\min = 0, \theta\max = 2\pi, \theta\text{step} = \frac{\pi}{24};$$

$$X\min = -6, X\max = 6, X\text{scl} = 1;$$

$$Y\min = -4, Y\max = 4, Y\text{scl} = 1$$

1. $r = 2$ Note: This graph is a circle.

2. $r = \dfrac{4}{\cos\theta}$ Note: This graph is a vertical line.

3. $r = 4\cos(6\theta)$

4. $r = 2 - 3\cos\theta$

5. $r = \sin 2\theta - 4(\cos 6\theta)^3$ Note: This graph is a star.

6. $r = 3(\cos 14\theta)^3$ Note: This graph is an explosion.

7. $r = 3\cos(64\theta)$

8. For this question, use the following window settings:

$$\theta\min = 0, \theta\max = 10\pi, \theta\text{step} = \frac{\pi}{24};$$

$$X\min = -6, X\max = 6, X\text{scl} = 1;$$

$$Y\min = -4, Y\max = 4, Y\text{scl} = 1$$

$$r = 4\sqrt{\cos 2\theta}$$

Note: This graph is a **lemniscate.**

9. For this question, use the following window settings:

$$\theta\min = 0, \theta\max = 8\pi, \theta\text{step} = \frac{\pi}{24};$$

$$X\min = -9, X\max = 9, X\text{scl} = 1;$$

$$Y\min = -6, Y\max = 6, Y\text{scl} = 1$$

$$r = 0.25\theta$$

Note: This graph is a **spiral of Archimedes.**

10. For this question, use the following window settings:

$$\theta\min = 0, \theta\max = 8\pi, \theta\text{step} = \frac{\pi}{24};$$

$$X\min = -6, X\max = 6, X\text{scl} = 0;$$

$$Y\min = -4, Y\max = 4, Y\text{scl} = 0$$

$$r = 4\cos(10\cos\theta)$$

Note: This graph is a **fairy.**

Glossary

Acute angle   An angle between 0° and 90°.

Adjacent side   In a right triangle, the side between a given acute angle and the right angle.

Altitude   A perpendicular line from one vertex of a triangle to the opposite side, or to an extension of the opposite side.

Ambiguous case   The case that arises when the only specifications given for a proposed triangle is the lengths of two sides and a non-included angle of the triangle (SSA). Three situations can occur: (1) no triangle exists, (2) one unique triangle exists, or (3) two distinct triangles may satisfy the conditions given.

Amplitude   One-half the absolute difference between the maximum and minimum values of a periodic function whose range is bounded.

Angle   The geometric figure formed by two rays with a common endpoint.

Angle of depression   The angle between the horizontal and the line of sight from an object to an observer's eye when the object is positioned lower than the observer.

Angle of elevation   The angle between the horizontal and the line of sight from an object to an observer's eye when the object is positioned higher than the observer.

Angular speed   The angle through which a rotating object travels per unit of time.

Arc length   The length of the curve formed by an arc.

Arccosecant   Another name for the inverse cosecant function.

Arccotangent   Another name for the inverse cotangent function.

Arccosine   Another name for the inverse cosine function.

Arcsecant   Another name for the inverse secant function.

Arcsine   Another name for the inverse sine function.

Arctangent   Another name for the inverse tangent function.

Area of a sector   The area of a portion of a circle bordered by two radii and the intercepted arc; equal to $\frac{1}{2}r^2\theta$, where θ is given in radians.

Argument   The input variable of a function; the angle, often designated θ, associated with the polar representation of a complex number $z = r(\cos\theta + i\sin\theta)$.

Asymptote   A line that the graph of a function gets closer and closer to in at least one direction along the line.

Cardioid   A polar graph named for its resemblance to a heart.

Central angle   An angle formed at the center of a circle by two radii.

Cofunction identities   The following set of identities associated with any two complementary angles α and β:

$$\sin\beta = \cos(90° - \beta) = \cos\left(\frac{\pi}{2} - \beta\right) = \cos\alpha$$

$$\cos\beta = \sin(90° - \beta) = \sin\left(\frac{\pi}{2} - \beta\right) = \sin\alpha$$

$$\tan\beta = \cot(90° - \beta) = \cot\left(\frac{\pi}{2} - \beta\right) = \cot\alpha$$

$$\sec\beta = \csc(90° - \beta) = \csc\left(\frac{\pi}{2} - \beta\right) = \csc\alpha$$

$$\csc\beta = \sec(90° - \beta) = \sec\left(\frac{\pi}{2} - \beta\right) = \sec\alpha$$

$$\cot\beta = \tan(90° - \beta) = \tan\left(\frac{\pi}{2} - \beta\right) = \tan\alpha$$

Complementary angles   Two positive angles whose sum is 90° or $\frac{\pi}{2}$.

Complex number   An expression $a + bi$, where a and b are real numbers and $i = \sqrt{-1}$.

Conditional equations   Equations that are true only for particular values of the variable(s) for all functions involved.

Cosecant   The reciprocal of the sine function.

Cosine function   The x coordinate of a point on the unit circle corresponding to a given angle.

Cotangent   The reciprocal of the tangent function.

Coterminal angles   Angles that have the same initial and terminal sides.

Counterexample   An example that disproves a statement.

De Moivre's theorem   Formula for finding the nth power or nth roots of a complex number.

Degree   A unit of angular measure equal to $\frac{1}{360}$ of a full revolution of a circle.

Double-angle formulas   Special cases of the sum and differences formulas for sine, cosine, and tangent in which the two angles are equal.

Even/odd identities   The set of equalities: $\sin(-\theta) = -\sin\theta$, $\cos(-\theta) = \cos\theta$, $\tan(-\theta) = -\tan\theta$, $\sec(-\theta) = \sec\theta$, $\csc(-\theta) = -\csc\theta$, and $\cot(-\theta) = -\cot\theta$, where if $f(-x) = -f(x)$, the function is odd, and if $f(-x) = f(x)$, the function is even.

Half-angle formulas   Identities derived from double-angle cosine formulas used to determine half-angle values of trigonometric functions.

Hypotenuse   In a right triangle, the side opposite the right angle.

Identities   Equations that are true for all values of the variable(s) for which all functions involved are defined.

Initial point   The starting point of a force, which usually is represented by an arrow indicating its magnitude and direction.

Initial side   The side of an angle from which rotation begins.

Inverse trigonometric functions   The following set of trigonometric functions defined for restricted domains of the sine, cosine, tangent, secant, cosecant, and cotangent functions:

Inverse sine function, symbolized by $\sin^{-1}x$, with domain $-1 \le x \le 1$ and output a real number (or angle) in the interval $\left[-\dfrac{\pi}{2}, \dfrac{\pi}{2}\right]$.

Inverse cosine function, symbolized by $\cos^{-1}x$, with domain $-1 \le x \le 1$ and output a real number (or angle) in the interval $[0, \pi]$.

Inverse tangent function, symbolized by $\tan^{-1}x$, with domain $-\infty < x < \infty$ and output a real number (or angle) in the interval $\left(-\dfrac{\pi}{2}, \dfrac{\pi}{2}\right)$.

Inverse secant function, symbolized by $\sec^{-1}x$, defined such that $\sec^{-1}x = \cos^{-1}\left(\dfrac{1}{x}\right)$ is a real number (or angle) $\left(\ne \dfrac{\pi}{2}\right)$ in the interval $[0, \pi]$.

Inverse cosecant function, symbolized by $\csc^{-1}x$, defined such that $\csc^{-1}x = \sin^{-1}\left(\dfrac{1}{x}\right)$ is a real number (or angle) $(\ne 0)$ in the interval $\left[-\dfrac{\pi}{2}, \dfrac{\pi}{2}\right]$.

Inverse cotangent function, symbolized by $\cot^{-1}x$, such that $\cot^{-1}x = \tan^{-1}\left(\dfrac{1}{x}\right)$ is a real number (or angle) $(\ne 0)$ in the interval $\left(-\dfrac{\pi}{2}, \dfrac{\pi}{2}\right]$.

Law of Cosines   For any triangle ABC, the formulas:

$$a^2 = b^2 + c^2 - 2bc\cos A$$
$$b^2 = a^2 + c^2 - 2ac\cos B$$
$$c^2 = a^2 + b^2 - 2ab\cos C$$

Law of Sines   For any triangle ABC, the equations:

$$\frac{\sin A}{a} = \frac{\sin B}{b} = \frac{\sin C}{c}$$

Magnitude of a force   The amount of a force without regard to its direction.

Measure of an angle   The amount of rotation from the initial side to the terminal side of an angle.

Midline   The horizontal line halfway between a periodic function's maximum and minimum values.

Modulus   The absolute value of a complex number, the distance $\sqrt{x^2 + y^2}$ from the origin to the point (x, y).

Negative angle   An angle measured clockwise from the positive x-axis.

Oblique triangle   A triangle that contains no right angle.

Obtuse angle   An angle between 90° and 180°.

Opposite side   In a right triangle, the side opposite a given acute angle.

One-to-one function   A function for which every element of the function's range corresponds to exactly one element of its domain.

Parameter   A variable upon which both x and y are dependent.

Period   The smallest number p of a periodic function f such that $f(x + p) = f(x)$.

Periodic function   A function f for which there exists a number P such that $f(x + P) = f(x)$ for any value of x.

Phase shift   The horizontal displacement of a basic trigonometric function.

Polar axis   In the polar coordinate system, a fixed ray emanating from the origin.

Polar coordinates   In the polar coordinate system, the coordinates (r, θ) of a point, where θ indicates the angle of rotation from the polar axis and r represents the radius, or the distance of the point from the pole in the direction of θ.

Polar equation   An equation describing a curve in the polar coordinate system.

Polar form of a complex number   The form $z = r(\cos\theta + i\sin\theta)$.

Pole   Another name for the origin in the polar coordinate grid.

Positive angle   An angle measured counterclockwise from the positive x-axis.

Product-to-sum formulas   The following trigonometric identities that allow the writing of a product of trigonometric functions as a sum or difference of trigonometric functions:

$$\sin\theta\cos\varphi = \frac{\sin(\theta + \varphi) + \sin(\theta - \varphi)}{2}$$

$$\cos\theta\sin\varphi = \frac{\sin(\theta + \varphi) - \sin(\theta - \varphi)}{2}$$

$$\sin\theta\sin\varphi = -\frac{\cos(\theta + \varphi) - \cos(\theta - \varphi)}{2}$$

$$\cos\theta\cos\varphi = \frac{\cos(\theta + \varphi) + \cos(\theta - \varphi)}{2}$$

Pythagorean identities   The set of identities: $\sin^2\theta + \cos^2\theta = 1$, $\tan^2\theta + 1 = \sec^2\theta$, and $\cot^2\theta + 1 = \csc^2\theta$.

Pythagorean theorem   In a right triangle, the formula $c^2 = a^2 + b^2$, where c is the length of the hypotenuse and a and b are the lengths of the legs.

Quadrantal angle   An angle whose terminal side lies on an axis.

Quotient identities   The pair of identities: $\tan\theta = \dfrac{\sin\theta}{\cos\theta}$ and $\cot\theta = \dfrac{\cos\theta}{\sin\theta}$.

Radian   The measure of a central angle of a circle that intercepts an arc whose length equals the circle's radius.

Ray   A half-line beginning at an endpoint and extending indefinitely in one direction from that point.

r cisθ   An abbreviation for $r(\cos\theta + i\sin\theta)$.

Reciprocal identities   The identities: $\sec\theta = \dfrac{1}{\cos\theta}$, $\csc\theta = \dfrac{1}{\sin\theta}$,

$\cot\theta = \dfrac{1}{\tan\theta}$, $\sin\theta = \dfrac{1}{\csc\theta}$, $\cos\theta = \dfrac{1}{\sec\theta}$, and $\tan\theta = \dfrac{1}{\cot\theta}$.

Rectangular form of a complex number   The form $z = a + bi$.

Reduction formulas   Identities used to reduce the complexity of a trigonometric function.

Reference angle   For a non-quadrantal angle in standard position, the acute angle formed by the terminal side of the angle and the x-axis.

Resultant   The single force obtained when two or more forces act at a point concurrently.

Reflex angle   An angle between 180° and 360°.

Right angle   An angle whose measure is 90°.

Secant   The reciprocal of the cosine function.

Sine function   The y coordinate of a point on the unit circle corresponding to a given angle.

Sinusoidal function   Any function that can be expressed in the form $f(x) = A\sin(Bx - C) + D$ or $f(x) = A\cos(Bx - C) + D$.

Solving a triangle   Determining the measures of all three angles and the lengths of all three sides of a triangle.

Standard position   The position of an angle with the vertex at the origin and the initial side along the positive x-axis.

Straight angle   An angle whose measure is 180°.

Sum/difference formulas   The set of formulas:

$$\sin(\theta + \phi) = \sin\theta\cos\phi + \cos\theta\sin\phi$$
$$\sin(\theta - \phi) = \sin\theta\cos\phi - \cos\theta\sin\phi$$

$$\cos(\theta + \phi) = \cos\theta\cos\phi - \sin\theta\sin\phi$$
$$\cos(\theta - \phi) = \cos\theta\cos\phi + \sin\theta\sin\phi$$

$$\tan(\theta + \varphi) = \frac{\tan\theta + \tan\varphi}{1 - \tan\theta\tan\varphi}$$

$$\tan(\theta - \varphi) = \frac{\tan\theta - \tan\varphi}{1 + \tan\theta\tan\varphi}$$

Sum-to-product formulas   The following trigonometric identities that allow the writing of a sum of trigonometric functions as a product of trigonometric functions:

$$\sin\theta + \sin\varphi = 2\sin\left(\frac{\theta + \varphi}{2}\right)\cos\left(\frac{\theta - \varphi}{2}\right)$$

$$\sin\theta - \sin\varphi = 2\cos\left(\frac{\theta + \varphi}{2}\right)\sin\left(\frac{\theta - \varphi}{2}\right)$$

$$\cos\theta + \cos\varphi = 2\cos\left(\frac{\theta + \varphi}{2}\right)\cos\left(\frac{\theta - \varphi}{2}\right)$$

$$\cos\theta - \cos\varphi = -2\sin\left(\frac{\theta + \varphi}{2}\right)\sin\left(\frac{\theta - \varphi}{2}\right)$$

Supplementary angles Two positive angles whose sum is 180°.

Tangent The quotient of the sine and cosine functions.

Terminal point The endpoint of a force, which usually is represented by an arrow indicating its magnitude and direction.

Terminal side The side of an angle at which its rotation ends.

Transformation A new function that results when one or more basic attributes of an original function are modified.

Triangle Inequality theorem The theorem that states that the sum of the lengths of any two sides of a triangle is greater than the length of the third side.

Unit circle The circle, centered at the origin, with radius 1. The equation is $x^2 + y^2 = 1$.

Verifying an identity Transforming one side of an identity until it is identical to the other side.

Calculator Instructions for Trigonometry Using the TI-84 Plus

Because you are a serious student of mathematics, we assume you already have a graphing calculator and that you have mastered the calculator's elementary operation. In this appendix, you will see that a graphing calculator is an indispensable tool for working with trigonometry. To demonstrate basic trigonometry features of graphing calculators, we have elected to use the TI-84 Plus Silver Edition platform. If you have a different calculator, that's okay. Most graphing calculators will have trigonometry features like the ones shown here. Consult your user's guidebook for instructions.

General Usage

▶ Use the arrow keys in the upper-right corner of the keyboard to move the cursor around the screen.

▶ Use the ENTER key to evaluate expressions and to execute commands.

▶ Use the blue 2ND key to access the secondary options printed in blue above the keys.

For a 2ND key action, rather than showing the primary key function, this appendix shows the secondary option in brackets. For example, 2ND ^ is shown as 2ND [π].

▶ Use 2ND [QUIT] to exit a menu.

▶ Use 2ND [ANS] to recall the previous result.

Setting the Calculator to Degree or Radian Mode

▶ To set your calculator to radian mode: press MODE, scroll down to the third row, highlight RADIAN, and then press ENTER.

▶ To set your calculator to degree mode: press MODE, scroll down to the third row, use the right arrow key to highlight DEGREE, and then press ENTER.

Overriding Radian or Degree Mode

▶ To override radian mode, use 2ND [ANGLE] 1:° to enter a degree symbol after the entry and treat the result as a degree measure.

▶ To override degree mode, use 2ND [ANGLE] 3:ʳ to enter a radian symbol after the entry and treat the result as a radian measure.

Evaluating Trigonometric Functions

▶ Set the calculator to the desired mode (radians or degrees). To evaluate sine, cosine, or tangent, use their respective keyboard keys.

EXAMPLE

To evaluate $\sin\dfrac{\pi}{6}$, enter the keystroke sequence $\boxed{\text{SIN}}\boxed{\text{2ND}}\boxed{\pi}\boxed{\div}6\boxed{)}\ \boxed{\text{ENTER}}$ in radian mode. (Note: For convenience, numerical values are shown entered in full, rather than as separate keystrokes.) The display shows the following:

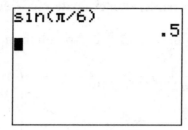

Be sure to enclose fractional angle measures in parentheses. Notice that the calculator automatically inserts a left parenthesis.

EXAMPLE

To evaluate cos 45°, enter the keystroke sequence $\boxed{\text{COS}}\ 45\ \boxed{\text{ENTER}}$ in degree mode. The display shows the following:

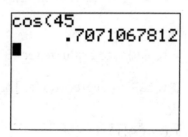

Notice that the closing parenthesis can be omitted.

EXAMPLE

To evaluate the tangent of the real number −1000, enter the keystroke sequence $\boxed{\text{TAN}}\boxed{(-)}1000\ \boxed{\text{ENTER}}$ in radian mode. The display shows the following:

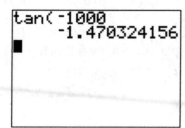

Be sure to use $\boxed{(-)}$, the negative key, not $\boxminus$, the minus key, to enter a negative value.

▶ The TI-84 Plus (like most calculators) does not have built-in secant, cosecant, and cotangent functions. Use the $\boxed{x^{-1}}$ key with the associated reciprocal identities to evaluate these functions.

EXAMPLE

▶ Because $\csc x = \dfrac{1}{\sin x} = (\sin x)^{-1}$, to evaluate $\csc\dfrac{\pi}{6}$, enter the keystroke sequence $\boxed{\text{SIN}}\boxed{\text{2ND}}[\pi]\boxed{\div}6\boxed{)}\boxed{x^{-1}}\boxed{\text{ENTER}}$ in radian mode. The display shows the following:

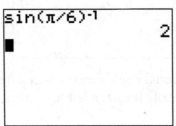

Do not omit the closing parenthesis for the function. Otherwise, your answer will be incorrect.

EXAMPLE

▶ Because $\sec\theta = \dfrac{1}{\cos\theta} = (\cos\theta)^{-1}$, to evaluate $\sec 45°$, enter the keystroke sequence $\boxed{\text{COS}}45\boxed{)}\boxed{x^{-1}}\boxed{\text{ENTER}}$ in degree mode. The display shows the following:

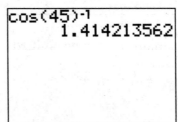

EXAMPLE

▶ Because $\cot x = \dfrac{1}{\tan x} = (\tan x)^{-1}$, to evaluate the cotangent of the real number -1000, enter the keystroke sequence $\boxed{\text{TAN}}\boxed{(-)}1000\boxed{)}\boxed{x^{-1}}\boxed{\text{ENTER}}$ in radian mode. The display shows the following:

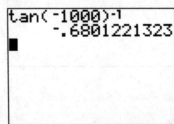

Determining Inverse Trigonometric Values

> If you want the answer in radians expressed in terms of π, use degree mode to obtain the answer in degrees, then multiply by $\frac{\pi}{180}$ to convert degrees to radians.

▶ Set the calculator to the desired mode. If you want the answer in radians, use radian mode.

▶ If you want the answer in degrees, use degree mode.

▶ Recall that the domains and ranges of the $\sin^{-1}$, $\cos^{-1}$, and $\tan^{-1}$ are as shown in the following table:

Inverse Function	Domain	Range
$\sin^{-1}$	$-1 \le x \le 1$	$-\dfrac{\pi}{2} \le y \le \dfrac{\pi}{2}$
$\cos^{-1}$	$-1 \le x \le 1$	$0 \le y \le \pi$
$\tan^{-1}$	$-\infty < x < \infty$	$-\dfrac{\pi}{2} < y < \dfrac{\pi}{2}$

▶ Calculators use the restricted domains and ranges of the inverse trigonometric functions. Therefore, for a given number, the calculator can return only one radian (or degree) value. Depending on the sign of the input number, the included quadrants in which an output may lie are quadrants I, II, or IV. See the table below for a summary of the included quadrants.

Inverse Function	Included Quadrant (Positive Input)	Included Quadrant (Negative Input)
$\sin^{-1}$	I	IV
$\cos^{-1}$	I	II
$\tan^{-1}$	I	IV

▶ To determine values of the inverse sine, inverse cosine, and inverse tangent functions, use their respective secondary function keys. For $\sin^{-1}$ use [2ND][$\sin^{-1}$], for $\cos^{-1}$ use [2ND][$\cos^{-1}$], and for $\tan^{-1}$ use [2ND][$\tan^{-1}$].

EXAMPLE

▶ To determine $\sin^{-1}(0.5)$ in degrees, enter the **keystroke** sequence
[2ND][$\sin^{-1}$].5[ENTER] in degree mode. The **display shows** the following:

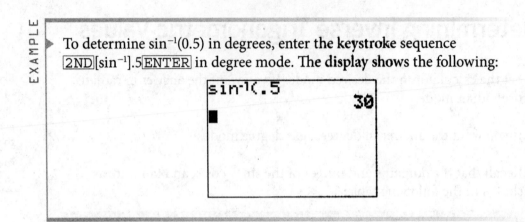

EXAMPLE

▶ To determine $\cos^{-1}(0.5)$ in degrees, enter the **keystroke** sequence
[2ND][$\cos^{-1}$].5[ENTER] in degree mode. The **display** shows the following:

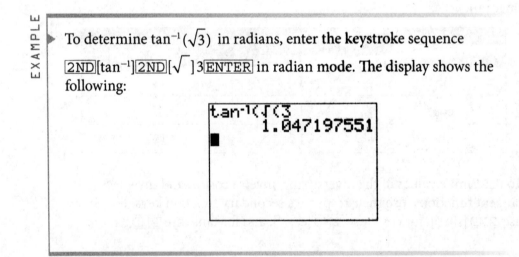

EXAMPLE

▶ To determine $\tan^{-1}(\sqrt{3})$ in radians, enter the **keystroke** sequence

[2ND][$\tan^{-1}$][2ND][$\sqrt{\ }$]3[ENTER] in radian **mode.** The display shows the
following:

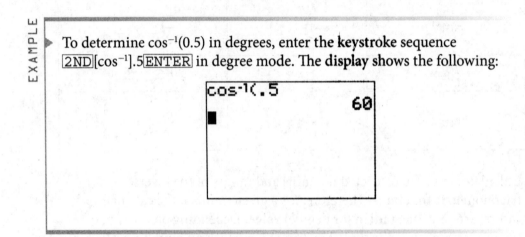

▶ To determine values of the inverse secant, **inverse cosecant,** and inverse
cotangent functions, use the reciprocal **identity's inverse** with the given
number's reciprocal as the input.

EXAMPLE

▶ Because $\sec^{-1} x = \cos^{-1}\left(\dfrac{1}{x}\right)$, to find $\sec^{-1}(2)$ in radians, enter the

keystroke sequence 2ND [cos⁻¹]2 x⁻¹ ENTER in radian mode. The display
shows the following:

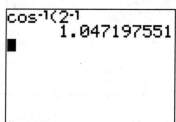

EXAMPLE

▶ Because $\csc^{-1} x = \sin^{-1}\left(\dfrac{1}{x}\right)$, to find $\csc^{-1}(1.414)$ in degrees, enter the

keystroke sequence 2ND [sin⁻¹]1.414 x⁻¹ ENTER in degree mode. The
display shows the following:

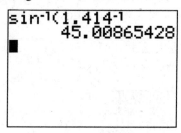

EXAMPLE

▶ Because $\cot^{-1} x = \tan^{-1}\left(\dfrac{1}{x}\right)$, to find $\cot^{-1}(-1.732)$ in degrees, enter the

keystroke sequence 2ND [tan⁻¹](−)1.732 x⁻¹ ENTER in degree mode. The
display shows the following:

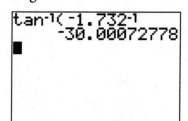

▶ Entering an input number that is not in an inverse function's domain
results in a calculator error message.

EXAMPLE

Entering the keystroke sequence 2ND[sin⁻¹]5 ENTER yields the following message:

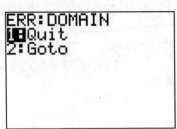

Graphing Polar Equations

▶ To graph a polar equation in radian mode, use the following steps: press MODE, scroll down to the third row, highlight RADIAN, and then press ENTER. Next, scroll to the fourth row, highlight POL, and then press ENTER. The display shows the following:

▶ Press WINDOW and set the values shown below. Note: These settings are typical, but can vary as needed.

$\theta \min = 0, \theta \max = 2\pi, \theta \text{step} = \dfrac{\pi}{24};$

$X \min = -6, X \max = 6, X\text{scl} = 1;$

$Y \min = -4, Y \max = 4, Y\text{scl} = 1$

A portion of the settings are displayed below.

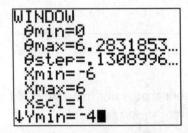

▶ Press Y= and enter an equation in the form $r = f(\theta)$. Use X,T,θ,n to enter θ into the equation. Then press GRAPH to plot the equation.

EXAMPLE

▶ **Polar Equation**　　　　　　　▶ **Polar Graph**

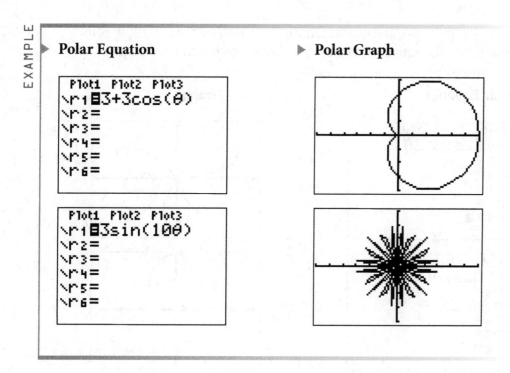

▶ To graph a polar equation in degree mode, press MODE and use the following settings:

▶ Press WINDOW and set the values shown below. Note: These settings are typical, but can vary as needed.

θmin $= 0$, θmax $= 360$, θstep $= 5$;

Xmin $= -2.35$, Xmax $= 2.35$, Xscl $= 1$;

Ymin $= -2$, Ymax $= 2$, Yscl $= 1$

A portion of the settings are displayed below.

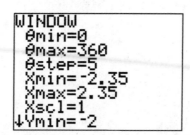

▶ Press $\boxed{Y=}$ and enter an equation in the form $r = f(\theta)$. Use $\boxed{X, T, \theta, n}$ to enter θ into the equation. Then press $\boxed{GRAPH}$ to plot the equation.

When needed, press $\boxed{ZOOM}$ 5: ZSquare to adjust the window and redraw the figure in a viewing window in which circular figures appear circular.

EXAMPLE

▶ **Polar Equation**

▶ **Polar Graph**

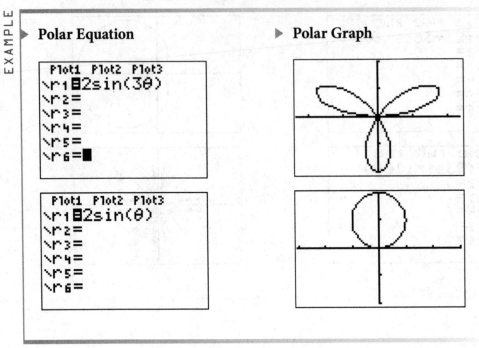

Trigonometric Identities

Reciprocal Identities

$$\csc\theta = \frac{1}{\sin\theta}$$

$$\sec\theta = \frac{1}{\cos\theta}$$

$$\cot\theta = \frac{1}{\tan\theta}$$

Cofunction Identities

$$\sin\left(\frac{\pi}{2} - \theta\right) = \cos\theta$$

$$\cos\left(\frac{\pi}{2} - \theta\right) = \sin\theta$$

$$\tan\left(\frac{\pi}{2} - \theta\right) = \cot\theta$$

Periodic Identities

$$\sin(\theta \pm 2\pi) = \sin\theta$$

$$\cos(\theta \pm 2\pi) = \cos\theta$$

$$\tan(\theta \pm \pi) = \tan\theta$$

Ratio Identities

$$\tan\theta = \frac{\sin\theta}{\cos\theta}$$

$$\cot\theta = \frac{\cos\theta}{\sin\theta}$$

Pythagorean Identities

$$\sin^2\theta + \cos^2\theta = 1$$

$$\tan^2\theta + 1 = \sec^2\theta$$

$$\cot^2\theta + 1 = \csc^2\theta$$

Addition/Subtraction Formulas

$$\sin(\theta \pm \varphi) = \sin\theta\cos\varphi \pm \cos\theta\sin\varphi$$

$$\cos(\theta \pm \varphi) = \cos\theta\cos\varphi \mp \sin\theta\sin\varphi$$

$$\tan(\theta \pm \varphi) = \frac{\tan\theta \pm \tan\varphi}{1 \mp \tan\theta\tan\varphi}$$

Product-to-Sum Formulas

$$\sin\theta\cos\varphi = \frac{\sin(\theta + \varphi) + \sin(\theta - \varphi)}{2}$$

$$\cos\theta\sin\varphi = \frac{\sin(\theta + \varphi) - \sin(\theta - \varphi)}{2}$$

Sum-to-Product Formulas

$$\sin\theta + \sin\varphi = 2\sin\left(\frac{\theta + \varphi}{2}\right)\cos\left(\frac{\theta - \varphi}{2}\right)$$

$$\sin\theta - \sin\varphi = 2\cos\left(\frac{\theta + \varphi}{2}\right)\sin\left(\frac{\theta - \varphi}{2}\right)$$

$$\sin\theta\sin\varphi = -\frac{\cos(\theta+\varphi) - \cos(\theta-\varphi)}{2} \qquad \cos\theta + \cos\varphi = 2\cos\left(\frac{\theta+\varphi}{2}\right)\cos\left(\frac{\theta-\varphi}{2}\right)$$

$$\cos\theta\cos\varphi = \frac{\cos(\theta+\varphi) + \cos(\theta-\varphi)}{2} \qquad \cos\theta - \cos\varphi = -2\sin\left(\frac{\theta+\varphi}{2}\right)\sin\left(\frac{\theta-\varphi}{2}\right)$$

Odd/Even Formulas	**Double-Angle Formulas**	**Half-Angle Formulas**

$$\sin(-\theta) = -\sin\theta \qquad \sin 2\theta = 2\sin\theta\cos\theta \qquad \sin\frac{\theta}{2} = \pm\sqrt{\frac{1-\cos\theta}{2}}$$

$$\cos(-\theta) = \cos\theta \qquad \cos 2\theta = \cos^2\theta - \sin^2\theta \qquad \cos\frac{\theta}{2} = \pm\sqrt{\frac{1+\cos\theta}{2}}$$

$$\tan(-\theta) = -\tan\theta \qquad \cos 2\theta = 2\cos^2\theta - 1 \qquad \tan\frac{\theta}{2} = \pm\sqrt{\frac{1-\cos\theta}{1+\cos\theta}}$$

$$\cos 2\theta = 1 - 2\sin^2\theta \qquad \tan\frac{\theta}{2} = \frac{\sin\theta}{1+\cos\theta}$$

$$\tan 2\theta = \frac{2\tan\theta}{1-\tan^2\theta} \qquad \tan\frac{\theta}{2} = \frac{1-\cos\theta}{\sin\theta}$$

$$\sin^2\theta = \frac{1-\cos 2\theta}{2}$$

$$\cos^2\theta = \frac{1+\cos 2\theta}{2}$$

The Complex Plane

A complex number of the form $z = a + bi$ can be put in a one-to-one correspondence with a point in the plane with coordinates (a,b). By this correspondence, any point in the plane can be considered the graph of a complex number. The plane in which complex numbers are represented is the **complex plane**. In a graph of the plane, the horizontal axis is the **real axis** and the vertical axis is the **imaginary axis**. The figure below illustrates the graph of the four complex numbers: $1 + i$, $-1 + 3i$, $-2 - 3i$, and $2 - i$, one in each quadrant.

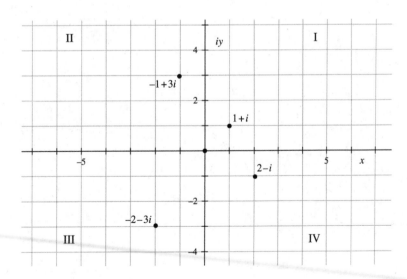

Answer Key

To the reader: Answers obtained using a calculator and rounded off may differ slightly from the rounded answers given in this answer key. Just so you know, it is the custom of the authors to not round off after intermediate steps, but to wait, if possible, until the last calculation to round off answers.

CHAPTER 1

Angles and Their Measure

EXERCISE 1-1

1. I
2. Negative *x*-axis
3. II
4. III
5. IV
6. II
7. Negative *y*-axis
8. I
9. III
10. II

11. True
12. False
13. False
14. False
15. True
16. False
17. True
18. True
19. False
20. True

EXERCISE 1-2

1. 50°	**11.** 60°
2. 74°	**12.** 158°
3. 2°	**13.** 99°
4. 36°	**14.** 90°
5. 43°	**15.** 79°
6. 69°	**16.** 120°
7. 4°	**17.** 14°
8. 29°	**18.** 177°
9. 70°	**19.** 37°
10. 77°	**20.** 162°

EXERCISE 1-3

1. 343°, IV	**11.** 50°
2. 20°, I	**12.** 50°
3. 106°, II	**13.** 15°
4. 259°, III	**14.** 85°
5. 346°, IV	**15.** 63°
6. 340°, IV	**16.** 40°
7. 204°, III	**17.** 17°
8. 67°, I	**18.** 88°
9. 46°, I	**19.** 15°
10. 182°, III	**20.** 35°

EXERCISE 1-4

1. $\dfrac{\pi}{6}$	**8.** 2π
2. $\dfrac{2\pi}{3}$	**9.** $\dfrac{3\pi}{4}$
3. $\dfrac{\pi}{4}$	**10.** $\dfrac{5\pi}{3}$
4. $\dfrac{\pi}{2}$	**11.** 540°
5. $\dfrac{3\pi}{2}$	**12.** 225°
6. π	**13.** 90°
	14. −45°
7. $\dfrac{\pi}{180}$	**15.** 150°
	16. 300°

17. 405°

18. 30°

19. −60°

20. 210°

21. 212.0°

22. 80.2°

23. 54.4°

24. 131.8°

25. 263.6°

26. $s = r\theta = (18 \text{ ft})\left(\dfrac{\pi}{3}\right) = 6\pi \text{ ft} \approx 18.8 \text{ ft}$

27. $s = r\theta = (12 \text{ m})\left(\dfrac{5\pi}{4}\right) = 15\pi \text{ m} \approx 47.1 \text{ m}$

28. $A = \dfrac{1}{2}r^2\theta = \dfrac{1}{2}(18 \text{ ft})^2\left(\dfrac{\pi}{3}\right) \approx 169.6 \text{ ft}^2$

29. $A = \dfrac{1}{2}r^2\theta = \dfrac{1}{2}(12 \text{ m})^2\left(\dfrac{5\pi}{4}\right) \approx 282.7 \text{ m}^2$

30. $\dfrac{(960)(2\pi)}{60 \text{ sec}} = 32\pi$ radians per second

CHAPTER 2

Concepts from Geometry

EXERCISE 2-1

1. Yes

2. No

3. Yes

4. No

5. Yes

6. 80°

7. 60°

8. 42°

9. 95°

10. 54°

11. 65°

12. 37°

13. 80°

14. 60°

15. 45°

16. Yes

17. Yes

18. No

19. Yes

20. Yes

EXERCISE 2-2

1. 13
2. 6
3. 24
4. 20
5. 37
6. 9
7. 45
8. 60
9. 16
10. 65

11. Yes
12. Yes
13. No
14. Yes
15. No
16. 24 feet
17. ≈ 33.3 feet
18. 50 feet
19. No. The door diagonal is only 8.5 feet.
20. ≈ 34.2 inches

CHAPTER 3

Right Triangle Trigonometry

EXERCISE 3-1

1. $\dfrac{10}{26}$

2. $\dfrac{24}{26}$

3. $\dfrac{26}{10}$

4. $\dfrac{24}{10}$

5. $\dfrac{10}{24}$

6. $\dfrac{26}{24}$

7. $\dfrac{24}{25}$

8. $\dfrac{25}{7}$

9. $\dfrac{25}{24}$

10. $\dfrac{7}{25}$

11. $\dfrac{24}{7}$

12. $\dfrac{7}{24}$

13. $\dfrac{9}{41}$

14. $\dfrac{35}{12}$

15. $\dfrac{11}{61}$

16. $\dfrac{6}{10}$

17. $\dfrac{40}{41}$

18. $\dfrac{35}{37}$

19. $\dfrac{4}{3}$

20. $\dfrac{13}{12}$

EXERCISE 3-2

1. $\cos 30° = \dfrac{b}{26}$ yields $b = 13\sqrt{3}$

2. $\sin 60° = \dfrac{a}{48}$ yields $a = 24\sqrt{3}$

3. $\sin 30° = \dfrac{20}{c}$ yields $c = 40$

4. $\sin 45° = \dfrac{a}{10}$ yields $a = 5\sqrt{2}$

5. $\tan 60° = \dfrac{9}{b}$ yields $b = 3\sqrt{3}$

6. $\sin 45° = \dfrac{20}{c}$ yields $c = 20\sqrt{2}$

7. $\cos 30° = \dfrac{b}{108}$ yields $b = 54\sqrt{3}$

8. $\sin 60° = \dfrac{a}{57}$ yields $a = \dfrac{57\sqrt{3}}{2}$

9. $\cos 45° = \dfrac{b}{16}$ yields $b = 8\sqrt{2}$

10. $\tan 30° = \dfrac{b}{12}$ yields $b = 4\sqrt{3}$

CHAPTER 4

General Right Triangles

EXERCISE 4-1

1. $B = 77°$, $b \approx 43.3$, $c \approx 44.5$
2. $B = 62°$, $a \approx 11.2$, $c \approx 23.8$
3. $A = 73°$, $a \approx 13.1$, $c \approx 13.7$
4. $B = 60°$, $a \approx 46.2$, $c \approx 92.4$
5. $B \approx 56.3°$, $A \approx 33.7°$, $c \approx 14.4$

6. $A \approx 43.9°$, $B \approx 46.1°$, $a \approx 17.3$
7. $A \approx 63.9°$, $B \approx 26.1°$, $a \approx 44.9$
8. $A \approx 36.9°$, $B \approx 53.1°$, $c = 5$
9. $A \approx 36.9°$, $B \approx 53.1°$, $c = 50$
10. $A = 65°$, $a \approx 17.2$, $c \approx 18.9$

EXERCISE 4-2

1. $(40 \text{ ft})(\sin 60°) \approx 34.6 \text{ ft}$

2. $\sin \theta = \dfrac{80}{140}$; $\theta \approx 34.8°$

3. $(150 \text{ ft})(\tan 52°) \approx 192.0 \text{ ft}$

4. $\tan 15° = \dfrac{x + 20}{100}$; $x \approx 6.8 \text{ ft}$

5. $\cos\left(\dfrac{\theta}{2}\right) = \dfrac{6}{10}$; $\theta \approx 106.3°$

6. $\tan \theta = \dfrac{6}{9}$; $\theta \approx 33.7°$

7. $(200 \text{ ft})(\sin 40°) \approx 128.6 \text{ ft}$

8. $\sin\left(\dfrac{\theta}{2}\right) = \dfrac{1.5}{5}$; $\theta \approx 34.9°$

9. $\tan A = \dfrac{15}{12}$; $A \approx 51.3°$

10. $2(4 \text{ ft})(\sin 34°) \approx 4.5 \text{ ft}$

CHAPTER 5

Oblique Triangles

EXERCISE 5-1

1. $b^2 = 6^2 + 16^2 - 2(6)(16)\cos 60°$; $b = 14$

2. $\cos\theta = \dfrac{2^2 + \left(2\sqrt{2}\right)^2 - \left(2\sqrt{5}\right)^2}{2(2)(2\sqrt{2})}$; $\theta = 135°$

3. $a^2 = 7^2 + 8^2 - 2(7)(8)\cos 30°$; $a \approx 4.0$

4. $\cos\theta = \dfrac{4^2 + 5^2 - 3^2}{2(4)(5)}$; $\theta \approx 36.9°$

5. $\sqrt{2.6^2 + 4.3^2 - 2(2.6)(4.3)\cos 140°} \approx 6.5$; thus, the resultant's magnitude ≈ 6.5 lb.

6. $\sqrt{30^2 + 20^2 - 2(30)(20)\cos 105°} \approx 40.1$; thus, the length of the guy wire ≈ 40.1 ft.

7. $\cos\theta = \dfrac{5.1^2 + 6.3^2 - 4.3^2}{2(5.1)(6.3)}$; $\theta \approx 42.7°$

8. $A \approx 95.7°$, $B \approx 50.7°$, $C \approx 33.6°$

9. $c \approx 3.4$, $A \approx 65.7°$, $B \approx 39.0°$

10. $b \approx 33.3$, $A \approx 44.9°$, $C \approx 25.1°$

11. $\sqrt{8^2 + 20^2 - 2(8)(20)\cos 18°} \approx 12.6$; thus, the distance ≈ 12.6 km.

12. $\sqrt{8^2 + 6^2 - 2(8)(6)\cos 113°} \approx 11.7$; thus, the length ≈ 11.7 cm.

13. $a = 7$, $b = 5$, $c = 4$; $A \approx 101.5°$, $B \approx 44.4°$, $C \approx 34.1°$

14. $\sqrt{500^2 + 600^2 - 2(500)(600)\cos 105°} \approx 874.8$; thus, the distance ≈ 874.8 km.

15. $\sqrt{1,500^2 + 2,000^2 - 2(1,500)(2,000)\cos 50°}$ $\approx 1,547.0$; thus, the length $\approx 1,547.0$ ft.

16. $\cos\theta = \dfrac{2.9^2 + 3.3^2 - 4.1^2}{2(2.9)(3.3)}$; $\theta \approx 82.5°$

17. True

18. False

19. False

20. True

EXERCISE 5-2

1. $c = \dfrac{\sqrt{2}\sin 75°}{\sin 45°} \approx 1.9$

2. $a = \dfrac{\sqrt{3}\sin 45°}{\sin 30°} \approx 2.4$

3. $b = \dfrac{4\sin 60°}{\sin 75°} \approx 3.6$

4. $b = \dfrac{20\sin 60°}{\sin 80°} \approx 17.6$

5. $a = 8$

6. $\sin\theta = \dfrac{10\sin 52°}{8}$; $\theta \approx 80.1°$

7. $\sin\theta = \dfrac{6\sin 30°}{2\sqrt{3}}$; $\theta = 60°$

8. $a = \dfrac{10\sin 75°}{\sin 60°} \approx 11.2$, $b = \dfrac{10\sin 45°}{\sin 60°} \approx 8.2$

9. $c = 5$

10. $b = \dfrac{15\sin 45°}{\sin 30°} \approx 21.2$

11. $c = \dfrac{15\sin 67°}{\sin 30°} \approx 27.6$

12. $c = \dfrac{8\sin 40°}{\sin 33°} \approx 9.4$

13. $\sin\theta = \dfrac{40\sin 60°}{50}$; $\theta \approx 43.9°$

14. $\dfrac{(50\text{ lb})\sin 42°}{\sin 113°} \approx 36$ lb

15. $\dfrac{(20 \text{ cm})\sin 20°}{\sin 120°} \approx 7.9 \text{ cm}$, $\dfrac{(20 \text{ cm})\sin 40°}{\sin 120°}$
$\approx 14.8 \text{ cm}$

16. $d = \dfrac{(1{,}500 \text{ ft})(\sin 16°)}{\sin 42°}(\sin 58°) \approx 524.0 \text{ ft}$

17. $\dfrac{(2{,}500 \text{ ft})\sin 78.5°}{\sin 54.2°} \approx 3{,}020.5 \text{ ft}$

18. $h = \dfrac{(400 \text{ ft})(\sin 73°)}{\sin 56.6°}(\sin 50.4°) \approx 353.0 \text{ ft}$

19. $\dfrac{(25 \text{ ft})(\sin 15°)}{\sin 116°}(\sin 41°) \approx 4.7 \text{ ft}$

20. False

EXERCISE 5-3

1. One

2. One

3. None

4. None

5. Two

6. One

7. One

8. None

9. None

10. One right triangle

11. Two

12. One right triangle

13. One

14. True

15. False

EXERCISE 5-4

1. $B = 55°$, $a \approx 64.2$, $b \approx 91.7$

2. No triangle

3. $A \approx 20.7°$, $B \approx 127.2°$, $C \approx 32.1°$

4. $b \approx 8.9$, $A = 85°$, $a \approx 9.4$

5. No triangle

6. $B \approx 11.6°$, $b \approx 136.5$, $C \approx 38.4°$

7. $B \approx 36.8°$

8. $b \approx 163.5$

9. $a \approx 134.8$

10. $b \approx 22.8$

11. $B \approx 77.7°$ or $102.3°$

12. $b \approx 8.5$

13. $c \approx 14.3$

14. $a \approx 105.8$

15. $c \approx 141.5$

16. $c \approx 71.7$

17. $A \approx 51.9°$

18. $B \approx 36.8°$ or $143.2°$

19. $b \approx 65.1$

20. $a \approx 0.8$

21. $\sqrt{150^2 + 120^2 - 2(150)(120)\cos 85°} \approx 183.7$; thus, the distance $\approx 183.7 \text{ yd}$

22. $\sqrt{120^2 + 80^2 - 2(120)(80)\cos 45°} \approx 85.0$; thus, the length $\approx 85.0 \text{ ft}$

23. $\dfrac{(110 \text{ ft})\sin 41.5°}{\sin 8.75°}\sin 39.75° \approx 306.4 \text{ ft}$

24. $s = \dfrac{30\sin 50°}{\sin 80°} \approx 23.3$, $h = \dfrac{30\sin 50°}{\sin 80°}\sin 50°$
≈ 17.9

25. $\dfrac{1{,}550 \text{ ft}}{\tan 38°} \approx 1{,}983.9$ ft

26. $\dfrac{(400 \text{ ft})\sin 37°}{\sin 95°} \approx 241.6$ ft

27. $\sqrt{3^2 + 3^2 - 2(3)(3)\cos 70°} \approx 3.4$; thus, the length ≈ 3.4 ft

28. $\sqrt{18^2 + 23^2 - 2(18)(23)\cos 130°} \approx 37.2$; thus, the resultant's magnitude ≈ 37.2 lb

29. $\dfrac{(24 \text{ m})(\sin 21°)}{\sin 14°}(\sin 35°) \approx 20.4$ m

30. $\dfrac{(50 \text{ yd})\sin 70°}{\sin 80°} \approx 47.7$ yd

EXERCISE 5-5

1. $\dfrac{1}{2}(25)(30)\sin 60° \approx 324.8$

2. $\dfrac{1}{2}(20)(15)\sin 150° \approx 75$

3. $\dfrac{1}{2}(18)(10)\sin 120° \approx 77.9$

4. $\dfrac{1}{2}(14)(17)\sin 50° \approx 91.2$

5. $\dfrac{1}{2}(24)(16) = 192$

6. $\dfrac{1}{2}(30)(16)\sin 26° \approx 105.2$

7. $\dfrac{1}{2}(15)(22)\sin 35° \approx 94.6$

8. $\dfrac{1}{2}(15)(20)\sin 40° \approx 96.4$

9. $\dfrac{1}{2}(68)(44)\sin 17° \approx 437.4$

10. $\dfrac{1}{2}(95)(30)\sin 28° \approx 669.0$

11. $\dfrac{1}{2}(25)(20)\sin 105° \approx 241.5$

12. $\dfrac{1}{2}(8)(15) = 60$

13. $\dfrac{1}{2}(8 \text{ ft})(6 \text{ ft})\sin 54° \approx 19.4 \text{ ft}^2$

14. $\dfrac{1}{2}(10 \text{ in})(10 \text{ in})\sin 60° \approx 43.3 \text{ in}^2$

15. $6 \cdot \dfrac{1}{2}(3 \text{ in})(3 \text{ in})\sin 60° \approx 23.4 \text{ in}^2$

CHAPTER 6

CHAPTER 6

Trigonometric Functions of Any Angle

EXERCISE 6-1

1. $\sin\theta = -\dfrac{1}{5}$, $\cos\theta = \dfrac{2\sqrt{6}}{5}$, $\tan\theta = -\dfrac{1}{2\sqrt{6}}$

2. $\sin\theta = -0.8$, $\cos\theta = 0.6$, $\tan\theta = -\dfrac{4}{3}$

3. $\sin\theta = \dfrac{\sqrt{7}}{4}$, $\cos\theta = -\dfrac{3}{4}$, $\tan\theta = -\dfrac{\sqrt{7}}{3}$

4. $\sin\theta = -\dfrac{15}{17}$, $\cos\theta = -\dfrac{8}{17}$, $\tan\theta = \dfrac{15}{8}$

5. $\sin\theta = \dfrac{1}{2}$, $\cos\theta = \dfrac{\sqrt{3}}{2}$, $\tan\theta = \dfrac{1}{\sqrt{3}}$

6. $\sec\theta = -\dfrac{4}{3}$, $\csc\theta = \dfrac{4}{\sqrt{7}}$, $\cot\theta = -\dfrac{3}{\sqrt{7}}$

7. $\sec\theta = -\dfrac{41}{9}$, $\csc\theta = -\dfrac{41}{40}$, $\cot\theta = \dfrac{9}{40}$

8. $\sec\theta = \dfrac{1}{0.8} = \dfrac{5}{4}$, $\csc\theta = \dfrac{1}{-0.6} = -\dfrac{5}{3}$,

$\cot\theta = \dfrac{0.8}{-0.6} = -\dfrac{4}{3}$

9. $\sec\theta = -\dfrac{13}{12}$, $\csc\theta = \dfrac{13}{5}$, $\cot\theta = -\dfrac{12}{5}$

10. $\sec\theta = \dfrac{25}{7}$, $\csc\theta = \dfrac{25}{24}$, $\cot\theta = \dfrac{7}{24}$

11. I, IV

12. II, IV

13. I, II

14. I, III

15. III, IV

16. III

17. II

18. I

19. III

20. II

21. $\sin\theta = -\dfrac{12}{13}$, $\tan\theta = \dfrac{12}{5}$

22. $\cos\theta = \dfrac{8}{10}$, $\tan\theta = -\dfrac{6}{8}$

23. $\sin\theta = \dfrac{40}{41}$, $\cos\theta = \dfrac{9}{41}$

24. $\sin\theta = \dfrac{7}{25}$, $\cos\theta = -\dfrac{24}{25}$

25. $\cos\theta = -\dfrac{8}{17}$, $\tan\theta = \dfrac{15}{8}$

EXERCISE 6-2

1. b

2. a

3. c

4. d

5. c

6. c

7. b

8. d

9. c

10. $\theta = 90° - 28°$; $\theta = 62°$

11. $\theta = 90° - 54°$; $\theta = 36°$

12. $\theta = \dfrac{\pi}{2} - \dfrac{\pi}{3}$; $\theta = \dfrac{\pi}{6}$

13. $\theta + 7° = 90° - 48°$; $\theta = 35°$

14. $\theta - 40° = 90° - 63°$; $\theta = 67°$

15. $2\theta + 5° = 90° - 15°$; $\theta = 35°$

16. $\dfrac{1}{2}\theta = \dfrac{\pi}{2} - \left(\dfrac{5}{2}\theta + \dfrac{\pi}{6}\right)$; $\theta = \dfrac{\pi}{9}$

17. $5\theta + \dfrac{\pi}{12} = \dfrac{\pi}{2} - \left(3\theta - \dfrac{\pi}{4}\right)$; $\theta = \dfrac{\pi}{12}$

18. $2\theta = 90° - \theta$; $\theta = 30°$

19. $7\theta + 15° = 90° - (3\theta + 40°)$; $\theta = 3.5°$

20. $\dfrac{1}{3}\theta + 20° = 90° - 50°$; $\theta = 60°$

EXERCISE 6-3

1. $\sin\theta = \dfrac{\sqrt{3}}{2}$, $\cos\theta = \dfrac{1}{2}$, $\tan\theta = \sqrt{3}$

2. $\sin\theta = -\dfrac{1}{2}$, $\cos\theta = \dfrac{\sqrt{3}}{2}$, $\tan\theta = -\dfrac{1}{\sqrt{3}}$

3. $\sin\theta = \dfrac{\sqrt{2}}{2}$, $\cos\theta = -\dfrac{\sqrt{2}}{2}$, $\tan\theta = -1$

4. $\sin\theta = \dfrac{5}{13}$, $\cos\theta = \dfrac{12}{13}$, $\tan\theta = \dfrac{5}{12}$

5. $\sin\theta = -\dfrac{\sqrt{3}}{2}$, $\cos\theta = -\dfrac{1}{2}$, $\tan\theta = \sqrt{3}$

6. $\sin\theta = -\dfrac{40}{41}$, $\cos\theta = -\dfrac{9}{41}$, $\tan\theta = \dfrac{40}{9}$

7. $\sin\theta = -\dfrac{\sqrt{2}}{2}$, $\cos\theta = \dfrac{\sqrt{2}}{2}$, $\tan\theta = -1$

8. $\sin\theta = \dfrac{5}{13}$, $\cos\theta = -\dfrac{12}{13}$, $\tan\theta = -\dfrac{5}{12}$

9. $\sin\theta = \dfrac{24}{25}$, $\cos\theta = -\dfrac{7}{25}$, $\tan\theta = -\dfrac{24}{7}$

10. $\sin\theta = -\dfrac{77}{85}$, $\cos\theta = \dfrac{36}{85}$, $\tan\theta = -\dfrac{77}{36}$

11. $\sec\theta = 2$, $\csc\theta = \dfrac{2}{\sqrt{3}}$, $\cot\theta = \dfrac{1}{\sqrt{3}}$

12. $\sec\theta = \dfrac{2}{\sqrt{3}}$, $\csc\theta = -2$, $\cot\theta = -\sqrt{3}$

13. $\sec\theta = -\dfrac{2}{\sqrt{2}}$, $\csc\theta = \dfrac{2}{\sqrt{2}}$, $\cot\theta = -1$

14. $\sec\theta = \dfrac{13}{12}$, $\csc\theta = \dfrac{13}{5}$, $\cot\theta = \dfrac{12}{5}$

15. $\sec\theta = -2$, $\csc\theta = -\dfrac{2}{\sqrt{3}}$, $\cot\theta = \dfrac{1}{\sqrt{3}}$

16. $\sec\theta = -\dfrac{41}{9}$, $\csc\theta = -\dfrac{41}{40}$, $\cot\theta = \dfrac{9}{40}$

17. $\sec\theta = \dfrac{2}{\sqrt{2}}$, $\csc\theta = -\dfrac{2}{\sqrt{2}}$, $\cot\theta = -1$

18. $\sec\theta = -\dfrac{13}{12}$, $\csc\theta = \dfrac{13}{5}$, $\cot\theta = -\dfrac{12}{5}$

19. $\sec\theta = -\dfrac{25}{7}$, $\csc\theta = \dfrac{25}{24}$, $\cot\theta = -\dfrac{7}{24}$

20. $\sec\theta = \dfrac{85}{36}$, $\csc\theta = -\dfrac{85}{77}$, $\cot\theta = -\dfrac{36}{77}$

EXERCISE 6-4

1. 0

2. 2

3. 6

4. 4

5. −9

6. 0

7. −5

8. 8

9. 16

10. 6

EXERCISE 6-5

1. $\tan 280°$

2. $\sin 210°$

3. $\cos \dfrac{11\pi}{6}$

4. $\sec \dfrac{3\pi}{4}$

5. $\cot \dfrac{5\pi}{3}$

6. $\csc \dfrac{5\pi}{3}$

7. $\sin 135°$

8. $\cos 270°$

9. $\tan \dfrac{2\pi}{3}$

10. $\sec 45°$

11. $\cos 30° = \dfrac{\sqrt{3}}{2}$

12. $\sec 45° = \sqrt{2}$

13. $\sin \dfrac{\pi}{3} = \dfrac{\sqrt{3}}{2}$

14. $\csc 90° = 1$

15. $\tan \dfrac{5\pi}{4} = 1$

16. $5\sqrt{3}\tan \dfrac{\pi}{6} = 5\sqrt{3}\left(\dfrac{1}{\sqrt{3}}\right) = 5$

17. $6\sin 30° - 2\cos 60° = 2$

18. $3\sqrt{2}\sin 45° + 2\sqrt{3}\cos 30° = 3\sqrt{2}\left(\dfrac{\sqrt{2}}{2}\right) + 2\sqrt{3}\left(\dfrac{\sqrt{3}}{2}\right) = 6$

19. $-\tan \dfrac{\pi}{4}\sin \dfrac{\pi}{6} = -\dfrac{1}{2}$

20. $\cos \dfrac{\pi}{3}\sin \dfrac{\pi}{3} = \dfrac{\sqrt{3}}{4}$

EXERCISE 6-6

1. $\sin(-10°) = -\sin 10°$

2. $\tan\left(-\dfrac{\pi}{12}\right) = -\tan\left(\dfrac{\pi}{12}\right)$

3. $\cot(-65°) = -\cot 65°$

4. $\sec(-48°) = \sec 48°$

5. $\cos\left(-\dfrac{2\pi}{9}\right) = \cos \dfrac{2\pi}{9}$

6. $\tan\theta = -\dfrac{1}{3}; \cot\theta = -3$

7. $\sin\theta = -\dfrac{11}{60}; \csc\theta = -\dfrac{60}{11}$

8. $\sec\theta = -\dfrac{25}{7}; \cos\theta = -\dfrac{7}{25}$

9. $\cot\theta = \dfrac{3}{4}; \tan\theta = \dfrac{4}{3}$

10. $\cos\theta = \dfrac{48}{73}; \sec\theta = \dfrac{73}{48}$

11. $\cos(-60°) = \cos 60° = \dfrac{1}{2}$

12. $\sin\left(-\dfrac{\pi}{6}\right) = -\sin\dfrac{\pi}{6} = -\dfrac{1}{2}$

13. $\tan(-45°) = -\tan 45° = -1$

14. $\sin(-30°) = -\sin 30° = -\dfrac{1}{2}$

15. $\cot\left(-\dfrac{\pi}{4}\right) = -\cot\dfrac{\pi}{4} = -1$

16. $3\sqrt{2}\sin(-45°) = -3\sqrt{2}\sin(45°) = -3$

17. $-4\sin 30° + 2\cos 30° = -2 + \sqrt{3}$

18. $-5\sqrt{2}\tan(45°) + 4\cos(60°) = -5\sqrt{2} + 2$

19. $-4\sin(60°)\cos(30°) = -3$

20. $-2\sqrt{3}\tan\left(\dfrac{\pi}{3}\right) - 10\tan\left(\dfrac{\pi}{4}\right) = -16$

EXERCISE 6-7

1. $-\cos 55°$

2. $-\cot\dfrac{\pi}{5}$

3. $-\sin 12°$

4. $-\csc 80°$

5. $-\tan 47°$

6. $-\sin 35°$

7. $-\cot 40°$

8. $-\cos\dfrac{\pi}{9}$

9. $\tan 70°$

10. $\csc 40°$

11. $-\cos 60°$

12. $-\tan 60°$

13. $\sin\dfrac{\pi}{4}$

14. $\cos\dfrac{\pi}{8}$

15. $\sin 45°$

16. 1

17. $-\dfrac{1}{\sqrt{3}}$

18. -1

19. $-\dfrac{1}{\sqrt{3}}$

20. $\dfrac{\sqrt{2}}{2}$

21. $-\sqrt{2}$

22. $-\sqrt{2}$

23. $2\sqrt{3}$

24. $\dfrac{\sqrt{2}}{4}$

25. $-\dfrac{1}{\sqrt{3}}$

CHAPTER 7

Trigonometric Identities

EXERCISE 7-1

1. variable(s)

2. 0

3. identical

4. counterexample

5. cannot

EXERCISE 7-2

1. $\dfrac{\sin^3\theta}{\cos^3\theta} = \left(\dfrac{\sin\theta}{\cos\theta}\right)^3 = \tan^3\theta$

2. $\dfrac{\sec\theta}{\tan\theta} = \dfrac{1}{\cancel{\cos\theta}} \cdot \dfrac{\cancel{\cos\theta}}{\sin\theta} = \dfrac{1}{\sin\theta} = \csc\theta$

3. $\sec\theta\cot\theta = \dfrac{1}{\cancel{\cos\theta}} \cdot \dfrac{\cancel{\cos\theta}}{\sin\theta} = \csc\theta$

4. $\csc\theta\tan\theta = \dfrac{1}{\cancel{\sin\theta}} \cdot \dfrac{\cancel{\sin\theta}}{\cos\theta} = \dfrac{1}{\cos\theta} = \sec\theta$

5. $\dfrac{\csc\theta}{\cot\theta} = \dfrac{1}{\cancel{\sin\theta}} \cdot \dfrac{\cancel{\sin\theta}}{\cos\theta} = \dfrac{1}{\cos\theta} = \sec\theta$

6. $\dfrac{\cot^2\theta}{\csc^2\theta} = \dfrac{\cos^2\theta}{\cancel{\sin^2\theta}} \cdot \dfrac{\cancel{\sin^2\theta}}{1} = \cos^2\theta$

7. $\dfrac{\sin^2 2\theta\cot 2\theta}{\cos 2\theta} = \dfrac{\sin^2 2\theta\,\cancel{\cos 2\theta}}{\cancel{\cos 2\theta}\,\sin 2\theta} = \sin 2\theta$

8. $\sin\theta\cot^2\theta\sec^2\theta = \sin\theta\dfrac{\cancel{\cos^2\theta}}{\sin^2\theta}\dfrac{1}{\cancel{\cos^2\theta}} = \dfrac{1}{\sin\theta}$

$\qquad = \csc\theta$

9. $\dfrac{\sec\theta}{\cot\theta} = \dfrac{\dfrac{1}{\cos\theta}}{\dfrac{\cos\theta}{\sin\theta}} = \dfrac{1}{\cos\theta}\dfrac{\sin\theta}{\cos\theta} = \dfrac{\sin\theta}{\cos^2\theta}$

10. $\dfrac{\cot\theta}{\csc\theta} = \dfrac{\dfrac{\cos\theta}{\sin\theta}}{\dfrac{1}{\sin\theta}} = \dfrac{\cos\theta}{\cancel{\sin\theta}}\dfrac{\cancel{\sin\theta}}{1} = \cos\theta$

11. $\dfrac{\cot^2\theta}{\sec^2\theta} = \dfrac{\dfrac{\cos^2\theta}{\sin^2\theta}}{\dfrac{1}{\cos^2\theta}} = \dfrac{\cos^2\theta}{\sin^2\theta}\dfrac{\cos^2\theta}{1} = \dfrac{\cos^4\theta}{\sin^2\theta}$

12. $\sec\theta + \tan\theta = \dfrac{1}{\cos\theta} + \dfrac{\sin\theta}{\cos\theta} = \dfrac{1+\sin\theta}{\cos\theta}$

13. False

14. True

15. False

EXERCISE 7-3

1. $\dfrac{1}{\sin\theta\cos\theta}$

2. $\dfrac{1}{\sin\theta\cos\theta}$

3. $\cot^2\theta + 1 = \dfrac{\cos^2\theta}{\sin^2\theta} + 1 = \dfrac{\cos^2\theta + \sin^2\theta}{\sin^2\theta}$
$= \dfrac{1}{\sin^2\theta} = \csc^2\theta$

4. $\csc\theta + \sin\theta = \dfrac{1}{\sin\theta} + \sin\theta = \dfrac{\sin^2\theta + 1}{\sin\theta}$

5. $1 + \tan^2\theta = \sec^2\theta = \dfrac{1}{\cos^2\theta}$

6. $\cos^4\theta - \sin^4\theta = (\cos^2\theta - \sin^2\theta)(\cos^2\theta + \sin^2\theta)$
$= \cos^2\theta - \sin^2\theta$

7. $\cos^2\theta(1 + \tan^2\theta) = \cos^2\theta\sec^2\theta = \cos^2\theta\left(\dfrac{1}{\cos^2\theta}\right)$
$= 1$

8. $\dfrac{\sin^2\theta}{1 - \cos\theta} = \dfrac{1 - \cos^2\theta}{1 - \cos\theta} = \dfrac{(1 - \cos\theta)(1 + \cos\theta)}{1 - \cos\theta}$
$= 1 + \cos\theta = \dfrac{\sec\theta}{\sec\theta} + \dfrac{1}{\sec\theta} = \dfrac{\sec\theta + 1}{\sec\theta} = \dfrac{1 + \sec\theta}{\sec\theta}$

9. $\dfrac{1}{1 - \sin\theta} + \dfrac{1}{1 + \sin\theta} = \dfrac{(1 + \sin\theta) + (1 - \sin\theta)}{1 - \sin^2\theta}$
$= \dfrac{2}{\cos^2\theta}$

10. $(\csc^2\theta - 1)(\sec^2\theta - 1) = \cot^2\theta\tan^2\theta$
$= \dfrac{1}{\tan^2\theta}\tan^2\theta = 1$

11. $\dfrac{1 + \tan^2\theta}{\csc^2\theta} = \dfrac{\sec^2\theta}{\csc^2\theta} = \dfrac{\dfrac{1}{\cos^2\theta}}{\dfrac{1}{\sin^2\theta}} = \dfrac{\sin^2\theta}{\cos^2\theta}$
$= \left(\dfrac{\sin\theta}{\cos\theta}\right)^2 = \tan^2\theta$

12. $\dfrac{\sin^2\theta}{\tan^2\theta - \sin^2\theta} = \dfrac{\sin^2\theta}{\sin^2\theta\left(\dfrac{1}{\cos^2\theta} - 1\right)} = \dfrac{\cos^2\theta}{1 - \cos^2\theta}$
$= \dfrac{\cos^2\theta}{\sin^2\theta} = \cot^2\theta$

13. $\cos^2\theta(1 + \tan^2\theta) = \cos^2\theta\sec^2\theta = \cos^2\theta\dfrac{1}{\cos^2\theta} = 1$

14. $\tan^2\theta + \sec^2\theta = (\sec^2\theta - 1) + \sec^2\theta$
$= 2\sec^2\theta - 1 = \dfrac{2}{\cos^2\theta} - 1$

15. $\dfrac{1 + \tan^2\theta}{\tan^2\theta} = \dfrac{1}{\tan^2\theta} + 1 = \cot^2\theta + 1 = \csc^2\theta$

16. False

17. True

18. True

19. False

20. True

EXERCISE 7-4

1. $\sin(60° + 45°) = \dfrac{\sqrt{6} + \sqrt{2}}{4}$

2. $\sin(45° + 150°) = \dfrac{\sqrt{2} - \sqrt{6}}{4}$

3. $\sin(30° + 45°) = \dfrac{\sqrt{6} + \sqrt{2}}{4}$

4. $\sin(30° + 135°) = \dfrac{\sqrt{6} - \sqrt{2}}{4}$

5. $\sin\left(\dfrac{\pi}{6} + \dfrac{\pi}{4}\right) = \dfrac{\sqrt{6} + \sqrt{2}}{4}$

6. $\sin\left(\dfrac{\pi}{3} + \dfrac{\pi}{4}\right) = \dfrac{\sqrt{6} + \sqrt{2}}{4}$

7. 0

8. $\dfrac{\sqrt{3}}{2}$

9. $\sin(20° + 160°) = \sin 180° = 0$

10. $\sin(110° - 80°) = \sin 30° = \dfrac{1}{2}$

11. $\sin 6\theta$

12. $\sin 3\theta$

13. $\sin 4\theta$

14. $\sin 2\theta$

15. $\sin \theta$

16. $\sin 4\theta$

17. $\sin 14\theta$

18. $\sin \theta$

19. $\sin 2\theta$

20. $\sin 2\theta$

21. $\sin(180° + \theta) = \sin 180° \cos\theta + \cos 180° \sin\theta$
$= 0 \cdot \cos\theta + (-1)\sin\theta = -\sin\theta$

22. $\sin(360° - \theta) = \sin 360° \cos\theta - \cos 360° \sin\theta$
$= 0 \cdot \cos\theta - (1)\sin\theta = -\sin\theta$

23. $\sin(90° - \theta) = \sin 90° \cos\theta - \cos 90° \sin\theta$
$= 1 \cdot \cos\theta - (0) \cdot \sin\theta = \cos\theta$

24. $\sin\left(\dfrac{3\pi}{2} + \theta\right) = \sin\dfrac{3\pi}{2}\cos\theta + \cos\dfrac{3\pi}{2}\sin\theta$
$= (-1)\cos\theta + (0)\sin\theta = -\cos\theta$

25. $\sin\left(\theta - \dfrac{\pi}{6}\right) = \sin\theta\cos\left(\dfrac{\pi}{6}\right) - \cos\theta\sin\left(\dfrac{\pi}{6}\right)$
$= \sin\theta\left(\dfrac{\sqrt{3}}{2}\right) - \cos\theta\left(\dfrac{1}{2}\right) = \dfrac{\sqrt{3}\sin\theta - \cos\theta}{2}$

EXERCISE 7-5

1. $\cos(60° + 45°) = \dfrac{\sqrt{2} - \sqrt{6}}{4}$

2. $\cos(150° + 45°) = -\dfrac{\sqrt{2} + \sqrt{6}}{4}$

3. $\cos\left(\dfrac{\pi}{6} + \dfrac{\pi}{4}\right) = \dfrac{\sqrt{6} - \sqrt{2}}{4}$

4. $\cos(135° + 30°) = -\dfrac{\sqrt{6} + \sqrt{2}}{4}$

5. $\dfrac{\sqrt{2} + \sqrt{6}}{4}$

6. -1

7. $-\dfrac{1}{2}$

8. $-\dfrac{1}{2}$

9. $\cos(130° - 70°) = \cos 60° = \dfrac{1}{2}$

10. $\cos(35° + 55°) = \cos 90° = 0$

11. $\cos 4\theta$

12. $\cos \theta$

13. $\cos 2\theta$

14. $\cos 2\theta$

15. $\cos\theta$

16. $\cos 4\theta$

17. $\cos 2\theta$

18. $\cos\theta$

19. $\cos 2\theta$

20. $\cos 2\theta$

21. $\cos(180° - \theta) = \cos 180° \cos\theta + \sin 180° \sin\theta$
$= (-1)\cos\theta + (0)\sin\theta = -\cos\theta$

22. $\cos(180° + \theta) = \cos 180° \cos\theta - \sin 180° \sin\theta$
$= (-1)\cos\theta - (0)\sin\theta = -\cos\theta$

23. $\cos(360° - \theta) = \cos 360° \cos\theta + \sin 360° \sin\theta$
$= (1)\cos\theta + (0)\sin\theta = \cos\theta$

24. $\cos\left(\dfrac{5\pi}{2} + \theta\right) = \cos\dfrac{5\pi}{2}\cos\theta - \sin\dfrac{5\pi}{2}\sin\theta$
$= (0)\cos\theta - (1)\sin\theta = -\sin\theta$

25. $\cos\left(\theta - \dfrac{\pi}{3}\right) = \cos\theta\cos\dfrac{\pi}{3} + \sin\theta\sin\dfrac{\pi}{3}$
$= \cos\theta\left(\dfrac{1}{2}\right) + \sin\theta\left(\dfrac{\sqrt{3}}{2}\right) = \dfrac{\cos\theta + \sqrt{3}\sin\theta}{2}$

EXERCISE 7-6

1. $\tan(150° + 45°) = \dfrac{\sqrt{3} - 1}{\sqrt{3} + 1}$

2. $\tan(30° + 45°) = \dfrac{\sqrt{3} + 1}{\sqrt{3} - 1}$

3. $\tan(60° + 45°) = \dfrac{\sqrt{3} + 1}{1 - \sqrt{3}}$

4. $\tan\left(\dfrac{\pi}{3} - \dfrac{\pi}{4}\right) = \dfrac{\sqrt{3} - 1}{\sqrt{3} + 1}$

5. $\tan(135° + 30°) = \dfrac{1 - \sqrt{3}}{\sqrt{3} + 1}$

6. $\tan(130° + 50°) = \tan 180° = 0$

7. $\tan(110° - 50°) = \tan 60° = \sqrt{3}$

8. $\tan(115° - 70°) = \tan 45° = 1$

9. $\tan(100° + 50°) = \tan 150° = -\dfrac{1}{\sqrt{3}}$

10. $\tan(95° + 40°) = \tan 135° = -1$

11. $\tan 9\theta$

12. $\tan 4\theta$

13. $\tan 14\theta$

14. $\tan\theta$

15. $\tan 2\theta$

16. $-\tan 60°$

17. $-\tan 45°$

18. $\tan(180° - \theta) = \dfrac{\tan 180° - \tan\theta}{1 + \tan 180° \tan\theta}$
$= \dfrac{0 - \tan\theta}{1 + 0 \cdot \tan\theta} = -\tan\theta$

19. $\tan 2\theta = \tan(\theta + \theta) = \dfrac{\tan\theta + \tan\theta}{1 - \tan^2\theta}$
$= \dfrac{2\tan\theta}{1 - \tan^2\theta}$

20. $\tan\left(\theta + \dfrac{\pi}{4}\right) = \dfrac{\tan\theta + \tan\dfrac{\pi}{4}}{1 - \tan\theta\tan\dfrac{\pi}{4}} = \dfrac{\tan\theta + 1}{1 - \tan\theta}$
$= \dfrac{1 + \tan\theta}{1 - \tan\theta}$

EXERCISE 7-7

1. $\sin 2\theta = \cos 30°$; $\sin 2\theta = \dfrac{\sqrt{3}}{2}$; $2\theta = 60°$ or $120°$

$\theta = 30°$ or $60°$

2. $-\tan 3\theta = -\tan 120°$; $3\theta = 120°$ or $300°$ (reject);

$\theta = 40°$

3. Case 1: $-\sin 5\theta = \sin(45° + 2\theta)$; $\sin(-5\theta)$

$= \sin(45° + 2\theta)$; $-5\theta = 45° + 2\theta$; $-7\theta = 45°$;

$7\theta = -45°$; $7\theta = 315°$; $\theta = 45°$

Case 2: $\sin(180° + 5\theta) = \sin(360° + 45° + 2\theta)$;

$180° + 5\theta = 360° + 45° + 2\theta$; $3\theta = 225°$; $\theta = 75°$

4. $\cos(2\theta + 90°) = \cos(180° - 20°)$; $\cos(2\theta + 90°)$

$= \cos(160°)$; $2\theta + 90° = 160°$ or $200°$; 2θ

$= 70°$ or $110°$; $\theta = 35°$ or $55°$

5. $\theta + 40° = 2\theta + 5°$; $\theta = 35°$

EXERCISE 7-8

1. $2\tan 4\theta$

2. $\cos 10\theta$

3. $2\cos 12\theta$

4. $\sin 4\theta$

5. $\tan 2\theta$

6. $\dfrac{1}{2}\sin\theta$

7. $\cos 6\theta$

8. $\dfrac{\sin 6\theta}{24}$

9. $\dfrac{3}{10}\tan 4\theta$

10. $-\cos 2\theta$

11. $-\dfrac{\cot 8\theta}{3}$

12. $\dfrac{\tan 4\theta}{2}$

13. $\cos^4\theta - \sin^4\theta = (\cos^2\theta - \sin^2\theta)(\cos^2\theta + \sin^2\theta)$
$= \cos 2\theta$

14. $\dfrac{3}{2}\tan^2 2\theta$

15. $\dfrac{1 + \cos 2\theta}{2} = \dfrac{1 + \cos^2\theta - \sin^2\theta}{2} = \dfrac{2\cos^2\theta}{2}$
$= \cos^2\theta$

16. $\sin 2\theta \csc\theta = \dfrac{2\sin\theta\cos\theta}{\sin\theta} = 2\cos\theta$

17. $\dfrac{2}{1 + \cos 2\theta} = \dfrac{2}{1 + \cos^2\theta - \sin^2\theta} = \dfrac{2}{2\cos^2\theta}$
$= \sec^2\theta$

18. $\dfrac{1 - \tan^2 2\theta}{2\tan 2\theta} = \dfrac{1}{\dfrac{2\tan 2\theta}{1 - \tan^2 2\theta}} = \dfrac{1}{\tan 4\theta} = \cot 4\theta$

19. $\dfrac{\cot\theta + \tan\theta}{2} = \dfrac{1}{2}\left[\dfrac{\cos\theta}{\sin\theta} + \dfrac{\sin\theta}{\cos\theta}\right]$

$= \dfrac{1}{2}\left[\dfrac{\cos^2\theta + \sin^2\theta}{\sin\theta\cos\theta}\right]$

$= \dfrac{1}{\sin 2\theta} = \csc 2\theta$

20. $4\cos^3\theta\sin\theta - 4\cos\theta\sin^3\theta$
$= 4\cos\theta\sin\theta[\cos^2\theta - \sin^2\theta]$
$= 2\sin 2\theta\cos 2\theta = \sin 4\theta$

21. $\dfrac{\tan 3\theta}{1 - \tan^2 3\theta} = \dfrac{1}{2}\left(\dfrac{2\tan 3\theta}{1 - \tan^2 3\theta}\right) = \dfrac{1}{2}\tan 6\theta$

22. $\cos 3\theta = \cos(2\theta + \theta) = \cos 2\theta\cos\theta - \sin 2\theta\sin\theta$
$= (2\cos^2\theta - 1)\cos\theta - 2\sin\theta\cos\theta\sin\theta$
$= (2\cos^2\theta - 1 - 2\sin^2\theta)\cos\theta$
$= (2\cos^2\theta - 1 - 2(1 - \cos^2\theta))\cos\theta$
$= 4\cos^3\theta - 3\cos\theta$

23. $\sin\theta = -\dfrac{5}{13}, \cos\theta = -\dfrac{12}{13}$;

$\sin 2\theta = \dfrac{120}{169}$ and $\cos 2\theta = \dfrac{119}{169}$

24. $\sin\theta = -\dfrac{7}{25}, \cos\theta = -\dfrac{24}{25}$;

$\sin 2\theta = \dfrac{336}{625}$ and $\cos 2\theta = \dfrac{527}{625}$

25. $\sin\theta = -\dfrac{8}{17}, \cos\theta = \dfrac{15}{17}$;

$\sin 2\theta = -\dfrac{240}{289}$ and $\cos 2\theta = \dfrac{161}{289}$

26 $\sin\theta = \dfrac{1}{\sqrt{2}}, \cos\theta = \dfrac{1}{\sqrt{2}}$;

$\sin 2\theta = 1$ and $\cos 2\theta = 0$

27. $\sin\theta = \dfrac{4}{5}, \cos\theta = -\dfrac{3}{5}$;

$\sin 2\theta = -\dfrac{24}{25}$ and $\cos 2\theta = -\dfrac{7}{25}$

28. True

29. True

30. False

EXERCISE 7-9

1. $\sin\dfrac{\pi}{12} = \sqrt{\dfrac{1-\cos\dfrac{\pi}{6}}{2}} = \sqrt{\dfrac{1-\dfrac{\sqrt{3}}{2}}{2}} = \dfrac{\sqrt{2-\sqrt{3}}}{2}$

2. $\tan\left(\dfrac{30°}{2}\right) = \dfrac{\sin 30°}{1+\cos 30°} = \dfrac{\dfrac{1}{2}}{1+\dfrac{\sqrt{3}}{2}} = \dfrac{1}{2+\sqrt{3}}$

3. $\cos\dfrac{7\pi}{12} = -\sqrt{\dfrac{1+\cos\dfrac{7\pi}{6}}{2}} = -\sqrt{\dfrac{1-\dfrac{\sqrt{3}}{2}}{2}}$

$= -\dfrac{\sqrt{2-\sqrt{3}}}{2}$

4. $\cos\left(\dfrac{30°}{2}\right) = \sqrt{\dfrac{1+\cos 30°}{2}} = \sqrt{\dfrac{1+\dfrac{\sqrt{3}}{2}}{2}}$

$= \dfrac{\sqrt{2+\sqrt{3}}}{2}$

5. $\tan\left(\dfrac{60°}{2}\right) = \dfrac{\sin 60°}{1+\cos 60°} = \dfrac{\dfrac{\sqrt{3}}{2}}{1+\dfrac{1}{2}} = \dfrac{\sqrt{3}}{3}$

6. $\tan\left(\dfrac{315°}{2}\right) = \dfrac{\sin 315°}{1+\cos 315°} = \dfrac{-\dfrac{\sqrt{2}}{2}}{1+\dfrac{\sqrt{2}}{2}}$

$= -\dfrac{\sqrt{2}}{2+\sqrt{2}}$

7. $\cos\left(\dfrac{135°}{2}\right) = \sqrt{\dfrac{1+\cos 135°}{2}} = \sqrt{\dfrac{1-\dfrac{\sqrt{2}}{2}}{2}}$

$= \dfrac{\sqrt{2-\sqrt{2}}}{2}$

8. $\sin^2\dfrac{\theta}{2} = \dfrac{1-\cos\theta}{2} = \dfrac{\cos\theta\left(\dfrac{1}{\cos\theta}-1\right)}{2}$

$= \dfrac{\sec\theta-1}{2\sec\theta}$

9. $\dfrac{1+\cos 2\theta}{1-\cos 2\theta} = \dfrac{\dfrac{1+\cos 2\theta}{2}}{\dfrac{1-\cos 2\theta}{2}} = \dfrac{\cos^2\theta}{\sin^2\theta} = \cot^2\theta$

10. $\sin\theta\tan\dfrac{\theta}{2} = \dfrac{\sin^2\theta}{1+\cos\theta} = \dfrac{1-\cos^2\theta}{1+\cos\theta} = 1-\cos\theta$

$= \dfrac{\tan\theta}{\tan\theta}(1-\cos\theta) = \dfrac{\tan\theta-\sin\theta}{\tan\theta} = \dfrac{\tan\theta-\sin\theta}{\sin\theta\sec\theta}$

EXERCISE 7-10

1. $2\sin\dfrac{11\theta}{2}\sin\dfrac{5\theta}{2}$

2. $2\cos\dfrac{\theta}{2}\sin\dfrac{\theta}{4}$

3. $2\cos\dfrac{13\theta}{2}\cos\dfrac{5\theta}{2}$

4. $2\sin 5\theta\cos\theta$

5. $2\cos 3\theta\cos 2\theta$

6. $2\cos 12°\sin 6°$

7. $2\cos 4\theta\cos 2\theta$

8. $\dfrac{\sin\theta+\sin\varphi}{\cos\theta+\cos\varphi}=\dfrac{2\left[\sin\left(\dfrac{\theta+\varphi}{2}\right)\right]\left[\cos\left(\dfrac{\theta-\varphi}{2}\right)\right]}{2\left[\cos\left(\dfrac{\theta+\varphi}{2}\right)\right]\left[\cos\left(\dfrac{\theta-\varphi}{2}\right)\right]}$

$=\tan\dfrac{1}{2}(\theta+\varphi)$

9. $\dfrac{\sin 2\theta}{\sin 7\theta}=\dfrac{2\sin 2\theta\cos 7\theta}{2\sin 7\theta\cos 7\theta}=\dfrac{2\left(\dfrac{1}{2}\right)\left[\sin 9\theta+\sin(-5\theta)\right]}{\sin 14\theta}$

$=\dfrac{\sin 9\theta-\sin 5\theta}{\sin 14\theta}$

10. $\dfrac{\sin 6\theta-\sin 4\theta}{\cos 6\theta+\cos 4\theta}=\dfrac{2\cos 5\theta\sin\theta}{2\cos 5\theta\cos\theta}=\tan\theta$

EXERCISE 7-11

1. $\dfrac{1}{2}[\sin 4\theta+\sin 2\theta]$

2. $\dfrac{1}{2}[\cos 8\theta+\cos 4\theta]$

3. $\dfrac{1}{2}[\sin 10\theta-\sin 2\theta]$

4. $\dfrac{1}{2}[\cos 4\theta-\cos 12\theta]$

5. $\dfrac{1}{2}[\sin 13\theta+\sin(-3\theta)]=\dfrac{1}{2}[\sin 13\theta-\sin 3\theta]$

6. $\dfrac{1}{2}[\sin 10\theta-\sin 4\theta]$

7. $\sin 45°\sin 15°=\dfrac{1}{2}[\cos(45°-15°)-\cos(45°+15°)]$

$=\dfrac{1}{2}[\cos(30°)-\cos(60°)]=\dfrac{1}{2}\left[\dfrac{\sqrt{3}}{2}-\dfrac{1}{2}\right]$

$=\dfrac{\sqrt{3}-1}{4}$

8. $4\cos 4\theta\cos 2\theta\cos\theta=4\left(\dfrac{1}{2}\right)[\cos 6\theta+\cos 2\theta]\cos\theta$

$=2\cos 6\theta\cos\theta+2\cos 2\theta\cos\theta$

$=2\left(\dfrac{1}{2}\right)[\cos 7\theta+\cos 5\theta]+2\cos 2\theta\cos\theta$

$=\cos 7\theta+\cos 5\theta+2\cos 2\theta\cos\theta$

9. $\cos(\theta+45°)\cos\theta=\dfrac{1}{2}[\cos(2\theta+45°)+\cos(45°)]$

$=\dfrac{1}{2}[\cos(2\theta)\cos(45°)-\sin(2\theta)\sin(45°)+\cos(45°)]$

$=\dfrac{\sqrt{2}}{4}[\cos(2\theta)-\sin(2\theta)+1]$

$=\dfrac{\sqrt{2}}{4}[2\cos^2\theta-1-2\sin\theta\cos\theta+1]$

$=\dfrac{\sqrt{2}}{2}\cos\theta[\cos\theta-\sin\theta]$

10. $\sin(\theta+45°)\sin\theta=\dfrac{1}{2}[\cos(45°)-\cos(2\theta+45°)]$

$=\dfrac{1}{2}\left[\dfrac{\sqrt{2}}{2}-(\cos 2\theta\cos(45°)-\sin 2\theta\sin(45°))\right]$

$=\dfrac{\sqrt{2}}{4}[1-\cos 2\theta+\sin 2\theta]$

CHAPTER 8

Trigonometric Functions of Real Numbers

EXERCISE 8-1

1. 0.841	**16.** 1.851
2. 0.913	**17.** 2.450
3. −0.866	**18.** 2
4. −0.827	**19.** 1.778
5. −0.707	**20.** −1.414
6. 0.540	**21.** 1.188
7. 0.408	**22.** 1.095
8. 0.5	**23.** −1.155
9. 0.562	**24.** −1.209
10. −0.707	**25.** −1.414
11. 1.557	**26.** 0.642
12. 2.237	**27.** 0.447
13. −1.732	**28.** −0.577
14. −1.470	**29.** −0.680
15. 1	**30.** 1

EXERCISE 8-2

1. $f(x + nP)$	**6.** d
2. maximum, minimum	**7.** b
3. b	**8.** b
4. b	**9.** b
5. b	**10.** d

CHAPTER 9

Graphs of the Sine Function

EXERCISE 9-1

1. $-1 \le y \le 1$

2. 2π

3. 1

4. -1

5. 1

6. $y = 0$

7. $0, \pi, 2\pi$

8. $\dfrac{\pi}{2}$

9. $\dfrac{3\pi}{2}$

10. $0 < x < \pi$

11. $\pi < x < 2\pi$

12. Increasing from 0 to $\dfrac{\pi}{2}$ and from $\dfrac{3\pi}{2}$ to 2π

13. Decreasing from $\dfrac{\pi}{2}$ to $\dfrac{3\pi}{2}$

14. Yes

15. No

EXERCISE 9-2

1. 4

2. $\dfrac{1}{3}$

3. 1.5

4. 5

5. $\sqrt{2}$

6. 0.4

7. $\dfrac{4}{5}$

8. 10

9. 0.75

10. $\sqrt{3}$

11. $-4 \le y \le 4$

12. $-1.5 \le y \le 1.5$

13. $-0.4 \le y \le 0.4$

14. $-0.75 \le y \le 0.75$

15. $-\sqrt{3} \le y \le \sqrt{3}$

16. 10

17. $-\dfrac{4}{5}$

18. $0, \pi, 2\pi$

19. $\dfrac{3\pi}{2}$

20. $\dfrac{\pi}{2}$

21.

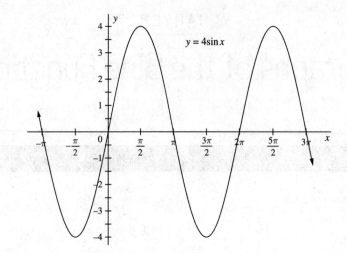

$y = 4\sin x$

22.

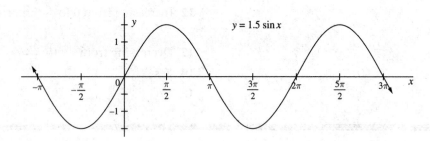

$y = 1.5 \sin x$

23.

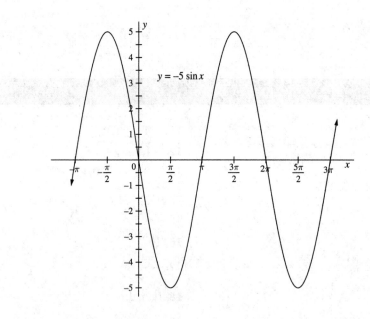

$y = -5 \sin x$

24.

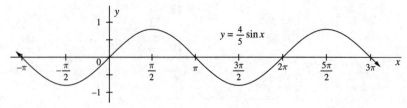

$y = \frac{4}{5}\sin x$

25.

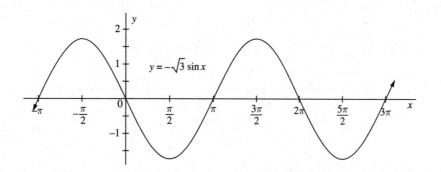

$$y = -\sqrt{3}\sin x$$

---

1. $\dfrac{\pi}{2}$

2. π

3. 2π

4. $\dfrac{\pi}{3}$

5. 6π

6. 4π

7. 2

8. 0.8π

9. $\dfrac{\pi}{4}$

10. 8π

11. $0, \dfrac{\pi}{4}, \dfrac{\pi}{2}, \dfrac{3\pi}{4}, \pi, \dfrac{5\pi}{4}, \dfrac{3\pi}{2}, \dfrac{7\pi}{4}, 2\pi$

12. $\dfrac{\pi}{8}, \dfrac{5\pi}{8}, \dfrac{9\pi}{8}, \dfrac{13\pi}{8}$

13. $\dfrac{3\pi}{8}, \dfrac{7\pi}{8}, \dfrac{11\pi}{8}, \dfrac{15\pi}{8}$

14. $0, 2\pi$

15. None

16. $y = -3\sin 2x$

17. $y = \sin 3x$

18. $y = 4\sin 2.5x$

19. $y = -\sqrt{2}\sin\left(\dfrac{x}{3}\right)$

20. $y = -\dfrac{4}{5}\sin\pi x$

21.

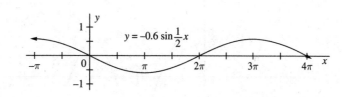

$$y = -0.6\sin\tfrac{1}{2}x$$

22.

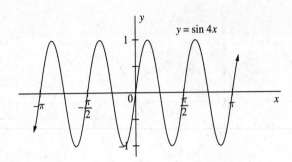

$y = \sin 4x$

23.

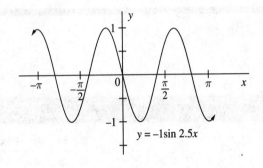

$y = -1\sin 2.5x$

24.

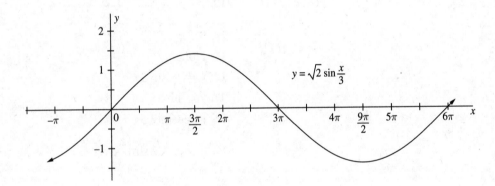

$y = \sqrt{2} \sin\frac{x}{3}$

25.

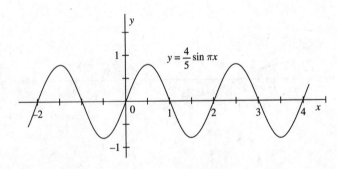

$y = \frac{4}{5}\sin \pi x$

EXERCISE 9-4

1. amplitude $= 5$, period $= 2\pi$, shift $= \dfrac{\pi}{4}$ right

2. amplitude $= 5$, period $= 2\pi$, shift $= \dfrac{\pi}{4}$ left

3. amplitude $= 1$, period $= \pi$, shift $= \dfrac{\pi}{6}$ left

4. amplitude $= \dfrac{1}{2}$, period $= \dfrac{2\pi}{3}$, shift $= \dfrac{\pi}{3}$ right

5. amplitude $= 3$, period $= \dfrac{2\pi}{3}$, shift $= \dfrac{\pi}{12}$ right

6. amplitude $= 6$, period $= \pi$, shift $= \dfrac{3}{2}$ right

7. amplitude $= 1$, period $= 2$, shift $= \dfrac{1}{3}$ right

8. amplitude $= \sqrt{2}$, period $= 4$, shift $= 2$ left

9. amplitude $= 0.4$, period $= 4\pi$, shift $= 6$ left

10. amplitude $= 2$, period $= 3\pi$, shift $= \dfrac{3}{2}$ right

11. $y = 3\sin\left(\dfrac{2x}{3} - \dfrac{\pi}{4}\right)$

12. $y = 2\sin\left(2x - \dfrac{\pi}{2}\right)$

13. $y = -\dfrac{1}{2}\sin\left(2x - 4\right)$

14. $y = -3\sin(2\pi x - \pi)$

15. $y = -\dfrac{3}{4}\sin\left(x + \dfrac{\pi}{3}\right)$

16. $\dfrac{3\pi}{8} \le x \le \dfrac{27\pi}{8}$

17. $\dfrac{\pi}{4} \le x \le \dfrac{5\pi}{4}$

18. $2 \le x \le 2 + \pi$

19. $\dfrac{1}{2} \le x \le \dfrac{3}{2}$

20. $-\dfrac{\pi}{3} \le x \le \dfrac{5\pi}{3}$

21.

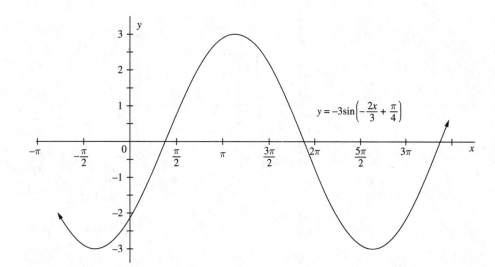

$y = -3\sin\left(-\dfrac{2x}{3} + \dfrac{\pi}{4}\right)$

22.

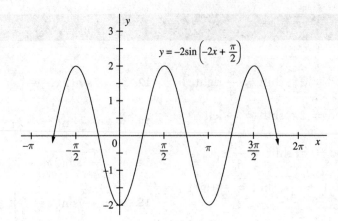

$$y = -2\sin\left(-2x + \frac{\pi}{2}\right)$$

23.

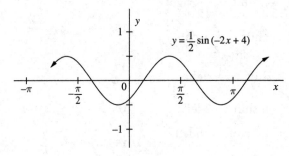

$$y = \frac{1}{2}\sin(-2x + 4)$$

24.

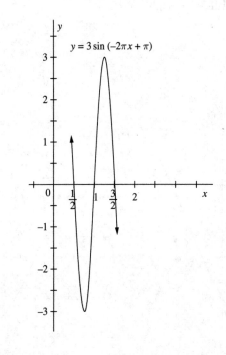

$$y = 3\sin(-2\pi x + \pi)$$

25.

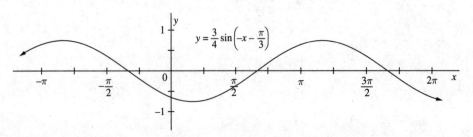

$$y = \frac{3}{4}\sin\left(-x - \frac{\pi}{3}\right)$$

EXERCISE 9-5

1. $\dfrac{\pi}{4}$ right, 1 downward

2. $\dfrac{\pi}{4}$ left, 6 upward

3. $\dfrac{\pi}{6}$ left, $\dfrac{1}{2}$ downward

4. $\dfrac{\pi}{3}$ right, $\sqrt{3}$ upward

5. $\dfrac{\pi}{12}$ right, 2.5 downward

6. $\dfrac{3}{2}$ right, $\dfrac{5}{4}$ upward

7. $\dfrac{1}{3}$ right, 21 upward

8. 2 left, 5.6 downward

9. 6 left, 9 downward

10. $\dfrac{3}{2}$ right, 45 upward

11. $y = \dfrac{5}{4}$

12. $y = 21$

13. $y = -5.6$

14. $y = -9$

15. $y = 45$

16.

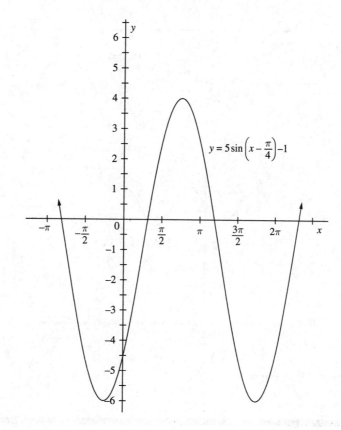

$$y = 5\sin\left(x - \frac{\pi}{4}\right) - 1$$

17.

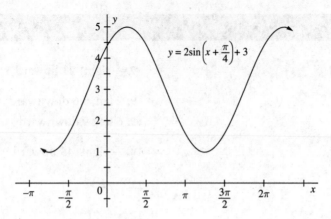

$$y = 2\sin\left(x + \frac{\pi}{4}\right) + 3$$

18.

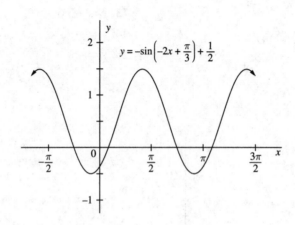

$$y = -\sin\left(-2x + \frac{\pi}{3}\right) + \frac{1}{2}$$

19.

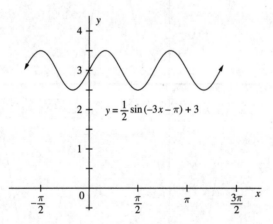

$$y = \frac{1}{2}\sin(-3x - \pi) + 3$$

20. period: $\dfrac{2\pi}{120\pi} = \dfrac{1}{60}$; max: 15 amperes

CHAPTER 10

Graphs of the Cosine Function

EXERCISE 10-1

1. $-1 \leq y \leq 1$

2. 2π

3. 1

4. -1

5. 1

6. $y = 0$

7. $\dfrac{\pi}{2}, \dfrac{3\pi}{2}$

8. $0, 2\pi$

9. π

10. $0 \leq x < \dfrac{\pi}{2}$ and $\dfrac{3\pi}{2} < x \leq 2\pi$

11. $\dfrac{\pi}{2} < x < \dfrac{3\pi}{2}$

12. Increasing from π to 2π

13. Decreasing from 0 to π

14. Yes

15. No

EXERCISE 10-2

1. $\dfrac{\pi}{4}$ right, 2 downward

2. $\dfrac{\pi}{4}$ left, 1 upward

3. $\dfrac{\pi}{12}$ left, $\dfrac{1}{2}$ upward

4. $\dfrac{\pi}{2}$ right, $\sqrt{5}$ upward

5. $\dfrac{\pi}{6}$ right, 0.4 downward

6. $\dfrac{2}{3}$ right, $\dfrac{5}{4}$ upward

7. $\dfrac{1}{3}$ right, 2 upward

8. 4 left, $\sqrt{2}$ upward

9. 9 left, 2 downward

10. $\dfrac{9}{4}$ right, 6 upward

11. $y = -2$

12. $y = 1$

13. $y = \dfrac{1}{2}$

14. $y = \sqrt{5}$

15. $y = -0.4$

16. $y = 3\cos 2x$

17. $y = -\cos\left(3x - \dfrac{\pi}{4}\right)$

18. $y = -4\cos\left(1.5x + 2\dfrac{\pi}{3}\right) + 2$

19. $y = \sqrt{2}\cos\left(\dfrac{x}{3} - \dfrac{\pi}{6}\right)$

20. $y = \dfrac{4}{5}\cos(\pi x + 4) - \dfrac{3}{2}$

21.

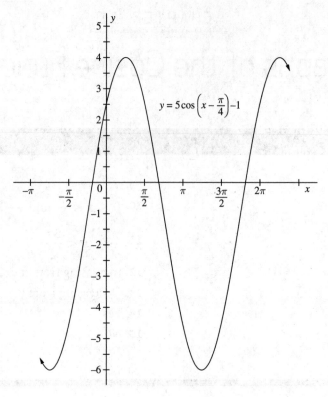

$$y = 5\cos\left(x - \frac{\pi}{4}\right) - 1$$

22.

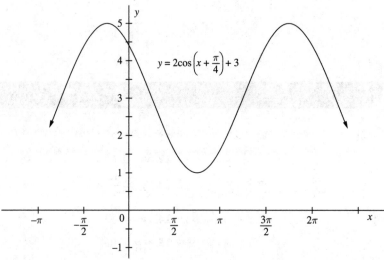

$$y = 2\cos\left(x + \frac{\pi}{4}\right) + 3$$

23.

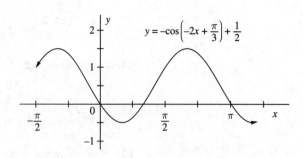

$$y = -\cos\left(-2x + \frac{\pi}{3}\right) + \frac{1}{2}$$

24.

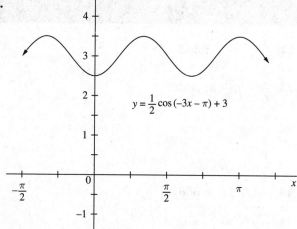

$$y = \frac{1}{2}\cos(-3x - \pi) + 3$$

25. period: 2; max: 10

CHAPTER 11

Graphs of the Tangent Function

EXERCISE 11-1

1. $\dfrac{\sin x}{\cos x}$

2. 0

3. R

4. $x = \dfrac{\pi}{2} + n\pi$

5. $n\pi$

6. π

7. undefined

8. False

9. True

10. True

EXERCISE 11-2

1. $|A| = 3$, period $= \pi$

2. $|A| = 5$, period $= \pi$

3. $|A| = 4$, period $= \dfrac{\pi}{2}$

4. $|A| = \dfrac{1}{2}$, period $= \dfrac{\pi}{2}$

5. $|A| = 0.6$, period $= \dfrac{\pi}{3}$

6. $\dfrac{2}{3}$ right, $\dfrac{5}{4}$ upward

7. $\dfrac{1}{3}$ right, 2 upward

8. 4 left, $\sqrt{2}$ upward

9. 9 left, 2 downward

10. $\dfrac{9}{4}$ right, 6 upward

11. $\dfrac{\pi}{4} + n\dfrac{\pi}{2}$, n odd

12. $-\dfrac{\pi}{4} + n\dfrac{\pi}{2}$, n odd

13. $-\dfrac{\pi}{12} + n\dfrac{\pi}{4}$, n odd

14. $\dfrac{\pi}{2} + n\dfrac{\pi}{4}$, n odd

15. $\dfrac{\pi}{6} + n\dfrac{\pi}{6}$, n odd

16. $\dfrac{2}{3} + n\dfrac{\pi}{6}$, n odd

17. $\dfrac{1}{3} + \dfrac{n}{2}$, n odd

18. $-4 + 2n$, n odd

19. $-9 + n\dfrac{3\pi}{2}$, n odd

20. $\dfrac{9}{4} + n\dfrac{3\pi}{8}$, n odd

21. $y = -3\tan 2x$

22. $y = \tan\left(3x - \dfrac{\pi}{4}\right)$

23. $y = 4\tan\left(1.5x + \dfrac{2\pi}{3}\right) + 2$

24. $y = -\sqrt{2}\tan\left(\dfrac{x}{3} - \dfrac{\pi}{6}\right)$

25. $y = -\tan\left(-x + \dfrac{\pi}{3}\right)$

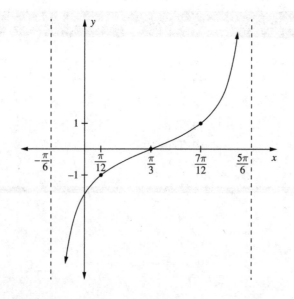

CHAPTER 12

Graphs of the Secant, Cosecant, and Cotangent Functions

1. $|A| = 3$, period $= 2\pi$

2. $|A| = 5$, period $= 2\pi$

3. $|A| = 4$, period $= \pi$

4. $|A| = \dfrac{1}{2}$, period $= \pi$

5. $|A| = 0.6$, period $= \dfrac{2\pi}{3}$

6. $\dfrac{2}{3}$ right, $\dfrac{5}{4}$ upward

7. $\dfrac{1}{3}$ right, 2 upward

8. 4 left, $\sqrt{2}$ upward

9. 9 left, 2 downward

10. $\dfrac{9}{4}$ right, 6 upward

11. $\dfrac{\pi}{4} + n\dfrac{\pi}{2}$, n odd

12. $-\dfrac{\pi}{4} + n\dfrac{\pi}{2}$, n odd

13. $-\dfrac{\pi}{12} + n\dfrac{\pi}{4}$, n odd

14. $\dfrac{\pi}{2} + n\dfrac{\pi}{4}$, n odd

15. $\dfrac{\pi}{6} + n\dfrac{\pi}{6}$, n odd

16. $\dfrac{2}{3} + n\dfrac{\pi}{6}$, n odd

17. $\dfrac{1}{3} + \dfrac{n}{2}$, n odd

18. $-4 + 2n$, n odd

19. $-9 + n\dfrac{3\pi}{2}$, n odd

20. $\dfrac{9}{4} + n\dfrac{3\pi}{8}$, n odd

21. $y = 3\sec 2x$

22. $y = -\sec\left(3x - \dfrac{\pi}{4}\right)$

23. $y = -4\sec\left(1.5x + \dfrac{2\pi}{3}\right) + 2$

24. $y = \sqrt{2}\sec\left(\dfrac{x}{3} - \dfrac{\pi}{6}\right)$

25. $y = \sec\left(x - \dfrac{\pi}{2}\right)$

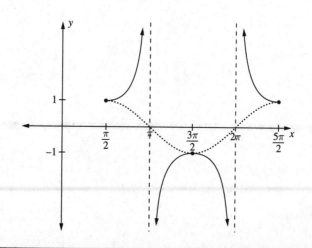

EXERCISE 12-2

1. $|A| = 3$, period $= 2\pi$
2. $|A| = 5$, period $= 2\pi$
3. $|A| = 4$, period $= \pi$
4. $|A| = \dfrac{1}{2}$, period $= \pi$
5. $|A| = 0.6$, period $= \dfrac{2\pi}{3}$
6. $\dfrac{2}{3}$ right, $\dfrac{5}{4}$ upward
7. $\dfrac{1}{3}$ right, 2 upward
8. 4 left, $\sqrt{2}$ upward
9. 9 left, 2 downward
10. $\dfrac{9}{4}$ right, 6 upward
11. $\dfrac{\pi}{4} + n\pi$
12. $-\dfrac{\pi}{4} + n\pi$
13. $-\dfrac{\pi}{12} + n\dfrac{\pi}{2}$

14. $\dfrac{\pi}{2} + n\dfrac{\pi}{2}$
15. $\dfrac{\pi}{6} + n\dfrac{\pi}{3}$
16. $\dfrac{2}{3} + n\dfrac{\pi}{3}$
17. $\dfrac{1}{3} + n$
18. $-4 + 4n$
19. $-9 + 3n\pi$
20. $\dfrac{9}{4} + 3n\dfrac{\pi}{4}$
21. $y = -3\csc 2x$
22. $y = \csc\left(3x - \dfrac{\pi}{4}\right)$
23. $y = 4\csc\left(1.5x + \dfrac{2\pi}{3}\right) + 2$
24. $y = -\sqrt{2}\csc\left(\dfrac{x}{3} - \dfrac{\pi}{6}\right)$
25. $y = \csc(-2x + \pi)$

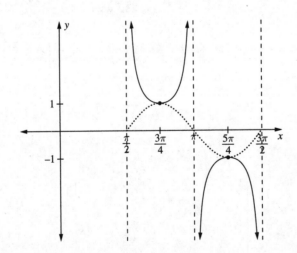

EXERCISE 12-3

1. $|A| = 3$, period $= \pi$

2. $|A| = 5$, period $= \pi$

3. $|A| = 4$, period $= \dfrac{\pi}{2}$

4. $|A| = \dfrac{1}{2}$, period $= \dfrac{\pi}{2}$

5. $|A| = 0.6$, period $= \dfrac{\pi}{3}$

6. $\dfrac{2}{3}$ right, $\dfrac{5}{4}$ upward

7. $\dfrac{1}{3}$ right, 2 upward

8. 4 left, $\sqrt{2}$ upward

9. 9 left, 2 downward

10. $\dfrac{9}{4}$ right, 6 upward

11. $\dfrac{\pi}{4} + n\pi$

12. $-\dfrac{\pi}{4} + n\pi$

13. $-\dfrac{\pi}{12} + n\dfrac{\pi}{2}$

14. $\dfrac{\pi}{2} + n\dfrac{\pi}{2}$

15. $\dfrac{\pi}{6} + n\dfrac{\pi}{3}$

16. $\dfrac{2}{3} + n\dfrac{\pi}{3}$

17. $\dfrac{1}{3} + n$

18. $-4 + 4n$

19. $-9 + 3n\pi$

20. $\dfrac{9}{4} + n\dfrac{3\pi}{4}$

21. $y = -3\cot 2x$

22. $y = \cot\left(3x - \dfrac{\pi}{4}\right)$

23. $y = 4\cot\left(1.5x + 2\dfrac{\pi}{3}\right) + 2$

24. $y = -\sqrt{2}\cot\left(\dfrac{x}{3} - \dfrac{\pi}{6}\right)$

25. $y = \cot\left(x + \dfrac{\pi}{6}\right)$

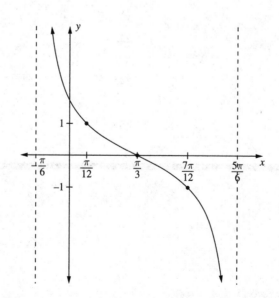

CHAPTER 13

Inverse Trigonometric Functions

EXERCISE 13-1

1. $\dfrac{\pi}{3}$

2. $\dfrac{\pi}{3}$

3. $\dfrac{\pi}{2}$

4. $-\dfrac{\pi}{4}$

5. $\dfrac{\pi}{6}$

6. $-\dfrac{\pi}{3}$

7. $-\dfrac{\pi}{6}$

8. $\dfrac{5\pi}{6}$

9. 0.4

10. 0.8

11. -1.1

12. -1.4

13. 0.5

14. 0.5

15. 1.5

16. $\sin\left(\sin^{-1}\dfrac{\sqrt{3}}{2}\right) = \sin\dfrac{\pi}{3} = \dfrac{\sqrt{3}}{2}$

17. $\tan\left(\cos^{-1}\dfrac{1}{2}\right) = \tan\dfrac{\pi}{3} = \sqrt{3}$

18. $\sin^{-1}\left(\sin\dfrac{3\pi}{4}\right) = \sin^{-1}\left(\dfrac{\sqrt{2}}{2}\right) = \dfrac{\pi}{4}$

19. $\cos^{-1}\left(\sin\dfrac{\pi}{6}\right) = \cos^{-1}\left(\dfrac{1}{2}\right) = \dfrac{\pi}{3}$

20. $\cos(\tan^{-1}\sqrt{3}) = \cos\left(\dfrac{\pi}{3}\right) = \dfrac{1}{2}$

21. $\dfrac{7}{25}$

22. $\dfrac{\sqrt{3}}{2}$

23. $\dfrac{4}{x}$

24. $\dfrac{3}{x}$

25. $\dfrac{4}{3}$

EXERCISE 13-2

1. $\dfrac{\pi}{6}$

2. $\dfrac{\pi}{6}$

3. $\dfrac{\pi}{4}$

4. $\dfrac{\pi}{4}$

5. $\dfrac{\pi}{3}$

6. 1.1

7. 0.3

8. 0.7

9. 0.8

10. 0.7

11. $\sec\left(\sec^{-1}\dfrac{2}{\sqrt{3}}\right) = \dfrac{2}{\sqrt{3}}$

12. $\cos(\cot^{-1}\sqrt{3}) = \cos\left(\dfrac{\pi}{6}\right) = \dfrac{\sqrt{3}}{2}$

13. $\csc^{-1}\left(\csc\dfrac{\pi}{4}\right) = \dfrac{\pi}{4}$

14. $\sec^{-1}\left(\csc\dfrac{\pi}{6}\right) = \sec^{-1}\left(\dfrac{2}{1}\right) = \dfrac{\pi}{3}$

15. $\tan\left(\sec^{-1}\dfrac{2}{\sqrt{3}}\right) = \tan\left(\dfrac{\pi}{6}\right) = \dfrac{1}{\sqrt{3}}$

16. $\dfrac{3}{x}$

17. $\dfrac{x}{\sqrt{x^2+9}}$

18. $\dfrac{x}{\sqrt{x^2+25}}$

19. $\dfrac{\sqrt{x^2+1}}{x}$

20. $\dfrac{\sqrt{x^2+4}}{2}$

CHAPTER 14

Solving Trigonometric Equations

EXERCISE 14-1

1. C

2. I

3. C

4. I

5. C

6. C

7. $2\cos\theta + 1 = 0$

$$\cos\theta = -\dfrac{1}{2}$$
$$\theta = 120° \text{ or } 240°$$

8. $\cos^2\theta - \sin^2\theta = \dfrac{\sqrt{3}}{2}$

$$\cos 2\theta = \dfrac{\sqrt{3}}{2}$$
$$2\theta = 30° \text{ or } 2\theta = 330°$$
$$\theta = 15° \text{ or } \theta = 165°$$

9.

$$\sin\theta\tan\frac{\theta}{2} = \frac{2-\sqrt{2}}{2}$$

$$\sin\theta\left(\frac{1-\cos\theta}{\sin\theta}\right) = \frac{2-\sqrt{2}}{2}$$

$$1-\cos\theta = \frac{2-\sqrt{2}}{2}$$

$$\cos\theta = \frac{\sqrt{2}}{2}$$

$$\theta = \frac{\pi}{4} \text{ or } \frac{7\pi}{4}$$

10. $\sec 3\theta = 2$

$$\cos 3\theta = \frac{1}{2}$$

$$3\theta = \frac{\pi}{3} \text{ or } \frac{5\pi}{3}$$

$$\theta = \frac{\pi}{9} \text{ or } \frac{5\pi}{9}$$

EXERCISE 14-2

Note: In this exercise set, some values are rejected because of the restriction $0 \le \theta \le \frac{\pi}{2}$ or $0° \le \theta \le 90°$.

1. $\cos 3\theta = 1$; $3\theta = 0$; $\theta = 0$

2. $\tan 4\theta = -\sqrt{3}$; $4\theta = \frac{2\pi}{3}$ or $\frac{5\pi}{3}$; $\theta = \frac{\pi}{6}$ or $\frac{5\pi}{12}$

3. $\sin 3\theta = 0$; $3\theta = 0$ or π; $\theta = 0$ or $\frac{\pi}{3}$

4. $\sin 2\theta = \cos 2\theta$; $\frac{\sin 2\theta}{\cos 2\theta} = 1$; $\tan 2\theta = 1$; $2\theta = \frac{\pi}{4}$ or $\frac{5\pi}{4}$(reject); $\theta = \frac{\pi}{8}$

5. $\cot 4\theta = -1$; $4\theta = \frac{3\pi}{4}$ or $\frac{7\pi}{4}$; $\theta = \frac{3\pi}{16}$ or $\frac{7\pi}{16}$

6. $\cos 5\theta = 1$; $5\theta = 0$ or 2π; $\theta = 0$ or $\frac{2\pi}{5}$

7. $\sin 5\theta = \frac{\sqrt{2}}{2}$; $5\theta = \frac{\pi}{4}, \frac{3\pi}{4}$, or $\frac{9\pi}{4}$; $\theta = \frac{\pi}{20}, \frac{3\pi}{20}$, or $\frac{9\pi}{20}$

8. $\sin\frac{\theta}{2} = \tan\frac{\theta}{2}$; $\sin\frac{\theta}{2} = \frac{\sin\frac{\theta}{2}}{\cos\frac{\theta}{2}}$; $\sin\frac{\theta}{2}\cos\frac{\theta}{2} = \sin\frac{\theta}{2}$; $\sin\frac{\theta}{2}\left(\cos\frac{\theta}{2}-1\right) = 0$; $\sin\frac{\theta}{2} = 0$, $\cos\frac{\theta}{2} = 1$; $\frac{\theta}{2} = 0$; $\theta = 0$

9. $\sin 2\theta - \cos\theta = 0$; $2\sin\theta\cos\theta - \cos\theta = 0$; $\cos\theta(2\sin\theta - 1) = 0$; $\cos\theta = 0$, $\sin\theta = \frac{1}{2}$; $\theta = \frac{\pi}{6}, \frac{5\pi}{6}$(reject) or $\frac{\pi}{2}$; $\theta = \frac{\pi}{6}$ or $\frac{\pi}{2}$

10. $\cos 2\theta + \cos\theta = 0$; $2\cos^2\theta - 1 + \cos\theta = 0$; $2\cos^2\theta + \cos\theta - 1 = 0$; $(2\cos\theta - 1)(\cos\theta + 1) = 0$; $\cos\theta = \frac{1}{2}$, $\cos\theta = -1$; $\theta = \frac{\pi}{3}$ or π(reject); $\theta = \frac{\pi}{3}$

11. $\sin 4\theta = \frac{1}{2}$; $4\theta = 30°$ or $150°$; $\theta = 7.5°$ or $37.5°$

12. $\sec 3\theta = \frac{2}{\sqrt{3}}$; $3\theta = 30°$ or $330°$ (reject); $\theta = 10°$

13. $\cot 5\theta = 1$; $5\theta = 45°, 225°$, or $405°$; $\theta = 9°, 45°$, or $81°$

14. $\cos 4\theta = \frac{1}{2}$; $4\theta = 60°$ or $300°$; $\theta = 15°$ or $75°$

15. $2\cos\theta - 1 = 0$; $\cos\theta = \frac{1}{2}$; $\theta = 60°$

16. $5\cos\theta - 2\sqrt{3} = \sqrt{3} - \cos\theta$; $6\cos\theta = 3\sqrt{3}$; $\cos\theta = \frac{\sqrt{3}}{2}$; $\theta = 30°$

17. $16\sin^2(2\theta) = 4$; $\sin^2(2\theta) = \frac{1}{4}$; $\sin 2\theta = \pm\frac{1}{2}$; $2\theta = 30°, 150°, 210°$(reject), $300°$(reject); $\theta = 15°$ or $75°$

18. $2\cos^2\theta - 2\sin^2\theta = \sqrt{3}$;
$\quad 2(\cos^2\theta - \sin^2\theta) = \sqrt{3}$;
$\quad 2(\cos 2\theta) = \sqrt{3}$; $\cos 2\theta = \dfrac{\sqrt{3}}{2}$;
$\quad 2\theta = 30°$ or $330°$ (reject); $\theta = 15°$

19. $4\cos\theta\sin\theta = \sqrt{2}$; $2(2\cos\theta\sin\theta) = \sqrt{2}$;
$\quad 2\sin 2\theta = \sqrt{2}$; $\sin 2\theta = \dfrac{\sqrt{2}}{2}$; $2\theta = 45°$ or $135°$;
$\quad \theta = 22.5°$ or $67.5°$

20. $\tan^2(2\theta) - 2\sqrt{3}\tan 2\theta = -3$;
$\quad \tan^2(2\theta) - 2\sqrt{3}\tan 2\theta + 3 = 0$;
$\quad \left(\tan 2\theta - \sqrt{3}\right)^2 = 0$; $\tan 2\theta = \sqrt{3}$;
$\quad 2\theta = 60°$ or $240°$ (reject); $\theta = 30°$

EXERCISE 14-3

1. $14.0°$, $166.0°$

2. $59.0°$, $301.0°$

3. $111.0°$, $291.0°$

4. $155.3°$, $204.7°$

5. $193.0°$, $347.0°$

6. $17.4°$, $197.4°$

7. $207.2°$, $332.8°$

8. $53.2°$, $306.8°$

9. $18.0°$, $198.0°$

10. $21.2°$, $338.8°$

CHAPTER 15

Trigonometric Form of a Complex Number

EXERCISE 15-1

1. $z = 2(\cos 60° + i\sin 60°)$

2. $z = \sqrt{2}(\cos 135° + i\sin 135°)$

3. $z = 2(\cos 210° + i\sin 210°)$

4. $z = 1(\cos 0° + i\sin 0°)$

5. $z = 4(\cos 60° + i\sin 60°)$

6. $z = 1(\cos 180° + i\sin 180°)$

7. $z = 1(\cos 90° + i\sin 90°)$

8. $z = 1(\cos 270° + i\sin 270°)$

9. $z = 2\sqrt{2}(\cos(-45°) + i\sin(-45°))$ or
$\quad 2\sqrt{2}(\cos 315° + i\sin 315°)$

10. $z = 2(\cos(-30°) + i\sin(-30°))$ or
$\quad 2(\cos 330° + i\sin 330°)$

EXERCISE 15-2

1. $z_1 z_2 = (2\text{cis}30°)(2\text{cis}60°) = 4\text{cis}90° = 4i$

2. $z_1 z_2 = 40(\text{cis}90°) = 40i$

3. $z_1 z_2 = 20\text{cis}240° = -10 - 10\sqrt{3}\,i$

4. $z_1 z_2 = 16.2\text{cis}\dfrac{11\pi}{15} \approx -10.8 + 12.0i$

5. $z_1 z_2 = 9\text{cis}109° \approx -2.9 + 8.5i$

6. $z_1 z_2 = (2\text{cis}(-30°))(2\text{cis}60°) = 4\text{cis}30°$
   $= 2\sqrt{3} + 2i$

7. $z_1 z_2 = (\sqrt{2}\text{cis}45°)(2\sqrt{2}\text{cis}45°) = 4\text{cis}90° = 4i$

8. $z_1 z_2 = (2\text{cis}30°)(2\text{cis}30°) = 4\text{cis}60° = 2 + 2\sqrt{3}i$

9. $z_1 z_2 = (1\text{cis}90°)(1\text{cis}(-90°)) = 1\text{cis}0° = 1$

10. $z_1 z_2 = (\sqrt{2}\text{cis}45°)(\sqrt{2}\text{cis}135°) = 2\text{cis}180° = -2$

11. $\dfrac{z_1}{z_2} = \dfrac{5}{2}\text{cis}30° = \dfrac{5\sqrt{3}}{4} + \dfrac{5}{4}i$

12. $\dfrac{z_1}{z_2} = \dfrac{5}{2}\text{cis}180° = -\dfrac{5}{2}$

13. $\dfrac{z_1}{z_2} = 5\text{cis}\left(-\dfrac{3\pi}{5}\right) \approx -1.6 - 4.8i$.

14. $\dfrac{z_1}{z_2} = 4\text{cis}55° \approx 2.3 + 3.3i$

15. $\dfrac{z_1}{z_2} = \dfrac{1}{2}\text{cis}(-90°) = -\dfrac{1}{2}i$

16. $V = IR = (4\text{cis}35°)(3\text{cis}25°) = 12\text{cis}60°$
    $= 6 + 6\sqrt{3}i$

17. $I = \dfrac{V}{R} = \dfrac{100(\text{cis}45°)}{25(\text{cis}60°)} = 4\text{cis}(-15°)$
    $= 4\text{cis}345°$

EXERCISE 15-3

1. $(-1 + i)^6 = (\sqrt{2}\text{cis}135°)^6 = 8\text{cis}810°$
   $= 8\text{cis}90° = 8i$

2. $[4(\cos 20° + i\sin 20°)]^3 = 64\text{cis}60°$
   $= 32 + 32\sqrt{3}i$

3. $(\text{cis}60°)^3 = \text{cis}180° = -1$

4. $[3(\cos 42° + i\sin 42°)]^5 = 243\text{cis}210°$
   $= -\dfrac{243\sqrt{3}}{2} - \dfrac{243}{2}i$

5. $[16(\cos 315° + i\sin 315°)]^{-2} = \dfrac{1}{256}\text{cis}(-630°)$
   $= \dfrac{1}{256}\text{cis}(90°) = \dfrac{1}{256}i$

6. $(3 + 3i)^6 = (3\sqrt{2}\text{cis}45°)^6 = 5{,}832\text{cis}270°$
   $= -5{,}832i$

7. $(4\cos 300° + 4i\sin 300°)^3 = 64\text{cis}900°$
   $= 64\text{cis}180° = -64$

8. $\left(\dfrac{\sqrt{2}}{2}\text{cis}135°\right)^8 = \dfrac{1}{16}\text{cis}1080° = \dfrac{1}{16}\text{cis}0° = \dfrac{1}{16}$

9. $(\sqrt{3} - i)^3 = (2\text{cis}(-30°))^3 = 8\text{cis}(-90°) = -8i$

10. $(4 + 3i)^5 = \left(5\text{cis}\left(\tan^{-1}\left(\dfrac{3}{4}\right)\right)\right)^5 \approx 3{,}125(-1.0 - 0.1i)$

11. $(1 + i)^{-4} = (\sqrt{2}\text{cis}45°)^{-4} = \dfrac{1}{4}\text{cis}(-180°) = -\dfrac{1}{4}$

12. $(\sqrt{3} + \sqrt{3}i)^6 = (\sqrt{6}\text{cis}45°)^6 = 216\text{cis}270° = -216i$

13. $(-3 - 3i)^{-3} = (3\sqrt{2}\text{cis}225°)^{-3} = \dfrac{1}{54\sqrt{2}}(\text{cis}(-675°))$
    $= \dfrac{1}{54\sqrt{2}}(\text{cis}45°) = \dfrac{1}{108} + \dfrac{1}{108}i$

14. $i^{-4} = \dfrac{1}{i^4} = \dfrac{1}{1} = 1$

15. $[5(\cos 15° + i\sin 15°)]^{-4} = \dfrac{1}{625}(\text{cis}(-60°))$
    $= \dfrac{1}{1{,}250} - \dfrac{\sqrt{3}}{1{,}250}i$

EXERCISE 15-4

1. $(\text{cis}\,330°)^{\frac{1}{2}}$: $\cos 165° + i\sin 165°$ and
$\cos 345° + i\sin 345°$

2. $[8(\cos 60° + i\sin 60°)]^{\frac{1}{3}}$: $2(\cos 20° + i\sin 20°)$,
$2(\cos 140° + i\sin 140°)$, $2(\cos 260° + i\sin 260°)$

3. $[81(\cos 30° + i\sin 30°)]^{\frac{1}{4}}$:
$3(\cos 7.5° + i\sin 7.5°)$,
$3(\cos 97.5° + i\sin 97.5°)$
$3(\cos 187.5° + i\sin 187.5°)$
$3(\cos 277.5° + i\sin 277.5°)$

4. Solve $z^3 + 1 = 0$ for all solutions. $r = 1$ and
$\theta = 180°$, Using De Moivre's theorem,

$$w_k = r^{\frac{1}{3}}\left(\cos\left(\frac{180° + k360°}{3}\right) + i\sin\left(\frac{180° + k360°}{3}\right)\right),$$
$k = 0,1,2$

You get the following three roots:

$w_0 = \cos 60° + i\sin 60° = \dfrac{1}{2} + i\dfrac{\sqrt{3}}{2}$,

$w_1 = \cos 180° + i\sin 180° = -1$, and

$w_2 = \cos 300° + i\sin 300° = \dfrac{1}{2} - i\dfrac{\sqrt{3}}{2}$.

(See the graph below.)

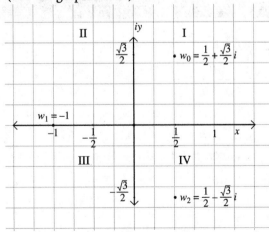

5. Solve $z^3 - 1 = 0$ for all solutions.

$r = 1$ and $\theta = 0°$. Using De Moivre's theorem,

$$w_k = r^{\frac{1}{3}}\left(\cos\left(\frac{0° + k360°}{3}\right) + i\sin\left(\frac{0° + k360°}{3}\right)\right),$$
$k = 0,1,2$

You get the following three roots:

$w_0 = \cos 0° + i\sin 0° = 1$,

$w_1 = \cos 120° + i\sin 120° = -\dfrac{1}{2} + i\dfrac{\sqrt{3}}{2}$, and

$w_2 = \cos 240° + i\sin 240° = -\dfrac{1}{2} - i\dfrac{\sqrt{3}}{2}$.

(See the graph below.)

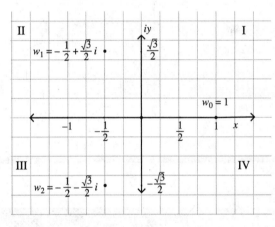

CHAPTER 16

Polar Coordinates

EXERCISE 16-1

1.

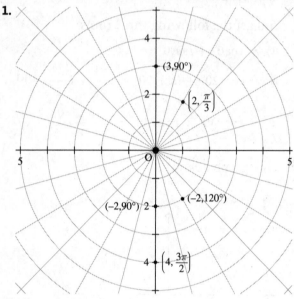

2.

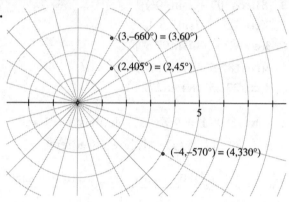

EXERCISE 16-2

1. $(2\sqrt{2}, 2\sqrt{2})$

2. $(3\sqrt{3}, 3)$

3. $(0, -5)$

4. $(3, -3\sqrt{3})$

5. $\left(-\dfrac{3}{2}, \dfrac{3\sqrt{3}}{2}\right)$

6. $(3, 0)$

7. $\left(4\sqrt{2}, \dfrac{\pi}{4}\right)$

8. $(5, 270°)$

9. $(1, 60°)$

10. $(2, 150°)$

11. $r\cos\theta = 5$

12. $r = \dfrac{\sin\theta}{4\cos^2\theta}$

13. $r = 5$

14. $r = \sqrt{\dfrac{6}{5\cos^2\theta + 1}}$

15. $r = 3\sin\theta$

16. $x^2 + y^2 = 64$

17. $x = 2$

18. $x^2 = 8y + 16$

19. $x^2 + y^2 = 3x$

20. $x^2 + y^2 - \sqrt{x^2 + y^2} - 2y = 0$

EXERCISE 16-3

1. Plot1 Plot2 Plot3
    \r₁■2
    \r₂=
    \r₃=
    \r₄=
    \r₅=
    \r₆=

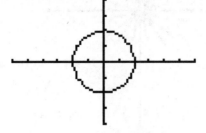

2. Plot1 Plot2 Plot3
    \r₁■4/cos(θ)
    \r₂=
    \r₃=
    \r₄=
    \r₅=
    \r₆=

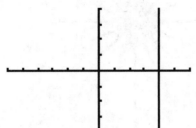

3. Plot1 Plot2 Plot3
    \r₁■4cos(6θ)
    \r₂=
    \r₃=
    \r₄=
    \r₅=
    \r₆=

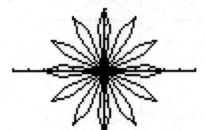

4. Plot1 Plot2 Plot3
    \r₁■2−3cos(θ)
    \r₂=
    \r₃=
    \r₄=
    \r₅=
    \r₆=

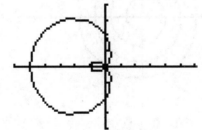

5. Plot1 Plot2 Plot3
    \r₁■sin(2θ)−4(co
    s(6θ))^3
    \r₂=
    \r₃=
    \r₄=
    \r₅=
    \r₆=

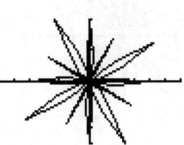

6.

Plot1 Plot2 Plot3

\r1■3cos(14θ)^3
\r2=
\r3=
\r4=
\r5=
\r6=

7.

Plot1 Plot2 Plot3

\r1■3cos(64θ)
\r2=
\r3=
\r4=
\r5=
\r6=

8.

Plot1 Plot2 Plot3

\r1■4√(cos(2θ))
\r2=
\r3=
\r4=
\r5=
\r6=

9.

Plot1 Plot2 Plot3

\r1■.25θ
\r2=
\r3=
\r4=
\r5=
\r6=

10.

Plot1 Plot2 Plot3

\r1■4cos(10cos(θ
))
\r2=
\r3=
\r4=
\r5=
\r6=

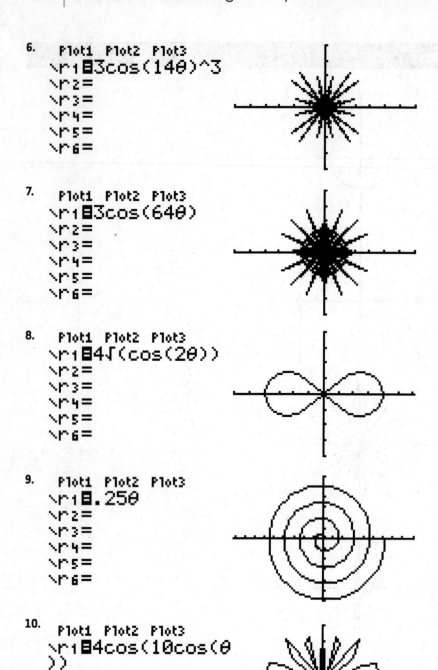